The In...

SY 0003817 2

D1313489

IBG STUDIES IN GEOGRAPHY

General Editors
Nigel Thrift and Mike Summerfield

IBG Studies in Geography are a range of stimulating texts which critically summarize the latest developments across the entire field of geography. Intended for students around the world, the series is published by Blackwell Publishers on behalf of the Institute of British Geographers.

Published

The European Community
Allan M. Williams

The Changing Geography of China
Frank Leeming

In preparation

Geography and Gender
Liz Bondi

Rural Geography
Paul Cloke

Service Industries in the World Economy
Peter Daniels

The Geography of Crime and Policing
Nick Fyfe

Fluvial Geography
Keith Richards

The Soviet Union in the Modern World
Denis Shaw

Tourism and Leisure
Gareth Shaw and Allan Williams

The Geography of Housing
Susan Smith

The Sources and Uses of Energy
John Soussan

Retail Restructuring
Neil Wrigley

DEBT AND DEVELOPMENT

Stuart Corbridge

BLACKWELL
Oxford UK & Cambridge USA

Copyright © Stuart Corbridge 1993

The right of Stuart Corbridge to be identified as author of this work has been asserted in accordance with the Copyright, Designs and Patents Act 1988.

First published 1993

Blackwell Publishers
108 Cowley Road,
Oxford OX4 1JF,
UK

238 Main Street, Suite 501,
Cambridge, Massachusetts 02142,
USA

British Library Cataloguing in Publication Data

A CIP catalogue record for this book is available from the British Library.

Library of Congress Cataloging-in-Publication Data

Corbridge, Stuart.
  Debt and development/Stuart Corbridge.
  p.   cm.—(IBG studies in geography)
  Includes bibliographical references.
  ISBN 0-631-17904-6.—ISBN 0-631-18138-5 (pbk.)
  1. Debts, External—Developing countries. 2 Economic development. I. Title. II. Series.
  HJ8899.C668 1992
  336.3'435'091724—dc20   92-17199   CIP

This book is printed on acid-free paper

Typeset in 11 on 13 point Plantin
by TecSet Ltd, Wallington, Surrey
Printed in Great Britain by T. J. Press Ltd, Padstow, Cornwall

Contents

Tables

Figures

Abbreviations

BIS	Bank for International Settlements
DRS	Debt Reporting System (of the World Bank)
ECLAC	Economic Commission for Latin America and the Caribbean
FDI	foreign direct investment
GATT	General Agreement on Tariffs and Trade
GDP	gross domestic product
GNP	gross national product
G-7	Group of Seven (Canada, France, Germany, Italy, Japan, United Kingdom and United States)
IBRD	International Bank for Reconstruction and Development
IDA	International Development Association
IECDI	Debt and International Finance Division of the World Bank's International Economics Department
IMF	International Monetary Fund
INT	total interest payments
LDCs	less developed countries
LIBOR	London Interbank Offered Rate
LICs	low-income countries
LINT	loan interest
MICs	middle-income countries
MYRAs	multiyear rescheduling arrangements
NICs	newly industrializing countries

ODA	official development assistance
OECD	Organization for Economic Co-operation and Development
OPEC	Organization of Petroleum Exporting Countries
RES	international reserves
SILIC	severely indebted low-income country
SIMIC	severely indebted middle-income country
SPA	Special Program for Africa
TDS	total debt service
TNCs	transnational corporations
UNCTAD	United Nations Conference on Trade and Development
XGS	exports of goods and services

Acknowledgements

Debt and Development is the product of many hours of classroom teaching, seminars, supervisions and informal discussions with friends and colleagues. For their help in different ways, I am grateful to: John Agnew and Jim and Nancy Duncan; Sue Roberts, Tim Lewington and Alfredo Robles; Chris Bramall, Mushtaq Khan and Alan Hughes; Ajit Singh and Geoff Hawthorn. Grateful thanks are also due to Graham Smith and Gerry Kearns, to Jenny Wyatt for preparing the artwork, and to Simon Prosser and Nigel Thrift for all their editorial help. Finally, my special thanks go to Sarah Jewitt. The book is dedicated to my daughter, Joanne.

The author and publishers gratefully acknowledge the following for permission to reproduce copyright material: Institute for International Economics (tables 3.2, 3.3 and 3.4); The Johns Hopkins University Press (table 2.3); Latin American Bureau, London (tables 2.6 and 5.1); The World Bank (tables 2.7 and 2.8, figures 2.1, 2.2, 2.5, 2.10, 3.1). The publishers apologize for any errors or omissions in the above list and would be grateful to be notified of any corrections that should be incorporated in the next edition or reprint of this book.

ONE

Introduction

This book has two main aims. It seeks first to embellish what I will call the standard narrative account of the developing countries' debt crisis. This is the account which is most akin to common sense and to popular conceptions of what the debt crisis is all about. It is an account which begins with the Organization of Petroleum Exporting Countries' (OPEC) oil-price shocks of 1973–4 and then traces through a trajectory which includes: petrodollar recycling through the Euromarkets; the second oil-price shocks of 1979–80 and the monetarist aftermath; a scissors crisis of rising real interest rates and declining real commodity prices; domestic economic misman-agement; the beginnings of an international debt crisis in Mexico in 1982 and its later generalization, in different forms, throughout most of Latin America and sub-Saharan Africa; structural adjust-ment programmes; and the Baker and Brady years.

Obviously, there is a lot more to this account that I have suggested thus far and there are good reasons for setting out this narrative at length, and in such a way that the whole of the 1980s debt crisis is considered. This is the aim of part I of the book, which comprises chapter 2. This chapter is written in a straightforward prose style that tries hard not to move away from the 'facts' in the direction of interpretation (always accepting that the two are intimately connected and that the standard narrative account itself is not unreflective). In this chapter, too, the student of the debt crisis is shown how to 'read' the debt crisis, in terms of the statistical sources which are available to trace its progress. Students these days are expected to show a competence in such things as econometrics and social theory; reading sets of economic statistics is a skill which is not always taught and which is sometimes neglected as a result. This book is old-fashioned in at least some of its teaching intentions.

Part II of the book adopts a different voice and concerns itself with three sets of interpretations of the developing countries' debt crisis: what I will call the system-stability, system-correction and system-instability perspectives on debt and development. This part of the book departs from the debt crisis in a narrow sense to consider how accounts of the debt crisis, and proposals for debt crisis management, might derive from certain underlying assumptions about the organization (and possible future organization) of international economic and political affairs.

Some of the arguments raised here touch upon development studies in a wider sense. We consider, for example, a series of arguments about markets and states in the developing world; about sovereignty, dependency and interdependency; about localism and globalism; and about development ethics in the short, medium and long terms. Chapters 3–5 allow a focus upon the debt crisis to serve as a means by which comments can be addressed to three traditions of development studies (counter-revolutionary, mainstream and radical) and to three traditions of social scientific thought (subjective preference theories, Keynesianism–pragmatism and radical political economy). In this manner the debt crisis takes on a symbolic importance as well as a practical significance. *Debt and Development* is a book about ideas – about ways of seeing – as much as it is a book about events and processes. The aim is not to suggest that a particular account of debt and development derives entirely from certain first principles (e.g. of Marxism or neo-classical economics); it is to suggest that many of the assumptions which are embedded in given accounts of 'debt' cannot properly be understood – and contested – without some knowledge of the relevant first principles.

A few words about each of chapters 3–5 are in order at this point. Chapter 3 presents a system-stability account of the debt crisis. The chapter begins by setting out some of the main assumptions and claims that underpin this perspective on debt and development. It would be tempting to say that these assumptions and claims are those of the New Right, but this is not quite the case. The New

Right often calls to mind a political ideology, such as Thatcherism, which links together a commitment to a free economy with a commitment to a strong state (Gamble, 1988). In the case of the system-stability perspective on the debt crisis there is little in the manner of a defence of the strong state. The system-stability perspective draws rather on various subjective preference theories – neo-classical, monetarist and Austrian economics, and a philosophy of libertarianism – to make the case for free economies and weak, or enabling, states. This is not a charter for conservatism. More to the point, these various subjective preference theories have helped to pave the way for a continuing counter-revolution in development theory and practice (to borrow a felicitous phrase of John Toye's: Toye, 1987). This counter-revolution in turn informs particular system-stability accounts of the debt crisis. These accounts emphasize the rationality and rectitude of economic transactions based in stable, competitive markets. They also highlight the responsibilities which actors, including countries, face for actions freely entered into. Profligate countries and banks are expected to bear the burdens of adjustment to a debt crisis very much of their own making. Debts should not be written down in a manner that rewards the profligate while punishing the prudent and the poor.

Chapter 4 provides a system-correction perspective on the debt crisis. This perspective has its roots in the twin assumptions and claims of Keynesianism and pragmatism, and these assumptions are set out in the second section of chapter 4. A third section outlines a prospectively Keynesian development studies, before section 4 examines how these wider assumptions and claims have informed a series of system-correction accounts of the debt crisis. The system-correction perspective is not anti-market in orientation, but it is sceptical of certain of the claims and assumptions made about competitive markets by subjective preference theorists. More so than a system-stability perspective, the system-correction account accepts that the developing countries' debt crisis may signal a crisis of global solvency (as opposed to localized crises of liquidity). Its proponents also suggest that no one set of actors is to blame for the debt crisis. Events happen in a manner which is not easy to predict and is even harder to control. This is particularly so when longstanding systems of economic regulation begin to break down. Actions to deal with the debt crisis should be correspondingly broad-based and pragmatic (section 5). Here, if you will, is the uncomfortable middle ground of development studies – a middle ground which resists abstract theories and sweeping conclusions and yet which remains the preferred terrain for most students of debt and development.

Chapter 5 presents a system-instability perspective on debt and development. Many of the assumptions and models which underlie this perspective derive from Marxism and radical political economy, and the bases of these traditions are set out in the second section of the chapter. Section 3 deals briefly with radical development studies. Sections 4 and 5 take up radical accounts of the origins and significance of the debt crisis (or crises), and the means by which these crises have been policed. The chapter is not confined to Marxist accounts of the debt crisis; hence the more general title under which it labours.

The book concludes with a brief discussion of the means by which these competing perspectives on debt and development might be judged. Chapter 6 also reflects upon the relationship between ideas and policies in economic and political affairs. *Debt and Development* takes seriously Keynes's claim that, 'the ideas of economists and political philosophers, both when they are right and when they are wrong, are more powerful than is commonly understood' (Keynes, 1936, p. 383). At the same time, the influence of ideas upon actions is not a direct one. The book concurs with Hirschman when he maintains that, 'it is difficult – and often ludicrous – to assign intellectual responsibility for actual policy decisions, let alone outcomes' (Hirschman 1981, 111). The influence of ideas upon actions is usually in the manner of 'indirect or recruitment effects' (ibid.), and this distinction is reflected in part in the organization of the book.

Part I of *Debt and Development* is about the debt crisis in 'fact' and with regard to existing policies for its management. Part II of the book is mainly about ideas; it is about the ways in which the debt crisis is theorized and produced in discourse, regardless of whether these ideas are put into practice in any direct fashion. At times, indeed, part II of the book moves in the direction of discourse analysis. It follows McCloskey in proposing that there is a rhetoric to economics and that much can be learned about economic affairs by attending to the linguistic tropes in and through which economic ideas are expressed and sometimes put into practice

(McCloskey 1986). One person's debt crisis may be another person's development crisis and another person's export crisis. Throughout the book, the reader is encouraged to think critically; indeed, each of chapters 3, 4 and 5 ends with some questions which might be put to the perspective previously outlined. The reader is asked to take little for granted (including the reader's guide to this literature, this book).

This brings me, conveniently, to some few words on the pros and cons of the approach I have taken in *Debt and Development*. In its favour, I hope, is a certain novelty of approach in a well-worked field, and a willingness to write chapters which play off one another as much as they act as complements to one another. Parts I and II of the book can be read independently of one another, although a good deal will be lost in the process. Any other merits which the book might have I leave to the reader to discover.

The book certainly has some pitfalls and I want to advertise these clearly to the reader at the outset. They have exercised me greatly and I have taken steps to minimize them.

A first pitfall concerns the regional coverage of *Debt and Development*. The book presents itself as a text on the *developing countries'* debt crisis, and it is for this reason that the reader will find little on the debt crisis in Eastern Europe or in some OECD countries (including the United States). These crises are mentioned only when they are relevant to some point of interpretation, as in chapter 5 where it is suggested that the developing countries' debt crisis is linked to a wider crisis of credit monies in post-Fordist capitalism. All this seems straightforward enough, but I am also conscious that the book moves uneasily between the debt crises which do affect the developing countries; between Latin America and sub-Saharan Africa, and parts of south-east Asia and elsewhere. Again, there are good reasons for this. Most accounts of the debt crisis confine themselves to the crises of private debt, which are centred in Latin America. Since this is in large part a book about interpretations of the debt crisis, it seems reasonable to follow this bias. Nevertheless, as a geographer I am reluctant to marginalize the non-Latin American countries in this way, and I have tried hard to build them into the narrative structure of chapter 2.

A second pitfall has to do with the idea of a paradigm, or a way of seeing. Part II of this book is presented as a considered intervention in a series of debates which go beyond the conventional boundaries of the debt crisis. These debates highlight the discursive construction of issues in development studies and they highlight the complexity of the relationships between ideas and events, models and policies. The problem, though, is in the implicit suggestion that there are three clear-cut perspectives on the debt and development nexus, and three only. This is not what I mean to suggest, notwithstanding the chapter headings in part II. I said earlier that I take Hirschman's point about the relationship between ideas and policies, and I emphasize that again now. By the same token, the paradigms advertised as chapters 3, 4 and 5 are not rigid and set in stone, and nor are all of the assumptions and models developed under each sign mutually exclusive. Capital flight is just one issue relevant to the debt crisis that crops up in all three chapters, albeit with subtle variations in the manner in which it is presented and interpreted. *Unnecessary* repetition is avoided as far as possible.

A third pitfall has to do with the writing of chapters 3–5. Some readers may feel that there is too much in the way of theoretical background in these chapters, while others may feel that there is not enough. On balance, I have opted to say more about a given intellectual tradition than is strictly necessary for an understanding of a particular perspective on the debt crisis. Thus Marxism is developed from first principles and with reference to the labour theory of value, when what are most relevant to chapter 5 are various Marxist and Marxissant theories of crisis formation and displacement under the rule of capital. There are three reasons for taking this tack. First, it saves the reader some of the trouble of looking outside the text for an account of the relevant first principles. Readers already familiar with these ideas and assumptions can skip section 2 (and possibly section 3) in each of chapters 3–5. Second, some readers will be familiar with Marxism, say, but not with subjective preference theories or with Keynesianism–pragmatism. For the sake of symmetry, if nothing else, it seemed sensible to write each chapter in a broadly similar fashion. Third, by outlining more of a given intellectual tradition than is necessary, I hope to enable the reader to ask more searching questions of any given account of the debt crisis.

A fourth pitfall concerns the relationship between intellectual paradigms and what we might call real scholars and real policy-makers: it is important to stress that the former do not imprison the latter. Although some individuals do belong to one camp and not to another, in many more instances this is not the case. A paper written by one scholar will often advance a series of claims which are more or less in keeping with a given paradigm, but add in others which are not. Scholars also change their tack as the years go by and as the issues they are studying refuse to stand still. People change their minds – especially people trained in the social sciences.

This book is offered in this spirit. It is offered in the spirit of Max Weber's remark that the true function of social science is to render problematic that which is conventionally self-evident. As such, it is concerned to challenge what seems obvious from a given perspective, but not to do this in a manner that does violence to the views of those it is seeking to represent. I hope the reader will agree that I have made every effort not to take views out of context and not to pigeon-hole authors (as opposed to particular discursive claims) when pigeon-holing is not called for. The reader must make a similar effort. The three paradigms set out in part II of the book are there for a reason and because they have a pedagogic value. They are not intended to serve as guides to rote-learning or as substitutes for a close empirical understanding of events and politics (see part I).

Finally, a few words of thanks. This book has grown out of lectures I have given on issues relating to debt and development over a number of years. Students of mine at Syracuse University and Cambridge University will be familiar with the approach that I have taken in this book, which is in keeping with my philosophy of teaching. What they might not so easily recognize is the contribution they have made to my own, imperfect processes of learning – both as a teacher and as a student of development issues. I am grateful to them all.

Part I

Describing the Debt Crisis –
a Standard Narrative Account

TWO

The Debt Crisis: a Standard Narrative Account

1 Introduction

Fleshing out a standard narrative account of the developing countries' debt crisis is not an easy task. Consider, for example, the following claims and discrepancies. In their account of *The Debt Squads: The US, the Banks and Latin America*, Branford and Kucinski (1988, p. 1) suggest that: 'To service its foreign debt, Latin America sent to the banks in the industrialized countries, in net terms, $159.1 billion from the end of 1981 to the end of 1986.' Peter Lindert, by contrast, claims that, 'most debtor countries have not been repaying significantly since 1982. The global net transfer from 97 developing-country debtors to creditors has been small as a share of debt outstanding. Nearly half the repayments received by private creditors have come on loans to just three "repayer" countries close to the United States [Mexico, Ecuador and Venezuela]' (Lindert 1989, 250). Meanwhile, P.-P. Kuczynski suggests that the 'implied resource transfer' from Latin America and the Caribbean, to all countries, totalled $86.8 billion between 1981 and 1986 (calculated from Kuczynski 1988, table 7.1).

These discrepancies are not the result of improper data analysis and presentation. They rather reflect differences in the reporting systems from which the data are sourced, differences in the manner in which current prices are adjusted for inflation and exchange-rate movements, differences in the scope of the creditor community referred to, differences in the types of transfer specified, and so on.

With these differences and discrepancies firmly in mind, this chapter has three main aims and comprises three parts and seven further sections. A first objective is to signal the range of data sources available to the student of the debt crisis (section 2), and to consider how stocks and flows of debt are most appropriately measured and understood (section 3). This first part may seem a little dry to some readers, but it is difficult to grasp what is at stake in the debates on the debt crisis without first understanding how debt statistics are defined, collected and presented. A second and larger purpose of the chapter is to provide an extended commentary on the dynamics of the developing countries' debt crisis. In the second part, the main point of focus is Latin America, as it is in most narrative accounts of the debt crisis. Section 4 considers the build-up to the debt crisis which exploded in August 1982. It considers how and why private forms of credit came increasingly to substitute for official development assistance and direct foreign investment (and not least in Latin America). It also considers the role of the banks and the changing structure of private credit flows to developing countries. Section 5 focuses upon the defaults of 1982 and 1983 and the reasons for this turn of events. The third part concludes the chapter with a lengthy examination of some of the debt management strategies that have been proposed and practised since 1982. Three strategies are considered in detail in sections 6–8: the containment/austerity strategy of 1982–5, the Baker strategy of adjustment with growth (1985–8), and the Brady/market-menu strategy which has been pursued since 1988/9. An assessment is offered of each strategy.

DATA AND DEFINITIONS

2   Data Sources

Data on developing country debt are reported in several publications. They are available in summary form in the *Statistical Yearbooks* of the United Nations, in the *Handbooks of International Trade and Development Statistics* published by the United Nations Conference on Trade and Development (UNCTAD), in the annual

Economic Surveys of Latin America and the Caribbean published by the UN's Economic Commission on Latin America and the Caribbean (ECLAC), and in various collections of national economic statistics and annual bank reports. The data there displayed, however, are data collected primarily by another agency: be it the International Monetary Fund (IMF) in the case of ECLAC, the Organization for Economic Co-operation and Development (OECD) in the case of UNCTAD, the World Bank in the case of the UN *Statistical Yearbook*, or the Bank for International Settlements (BIS). It is these four organizations which provide us with the most comprehensive and accessible statistics on the external debts of developing countries.

The debt reporting systems of the IMF, BIS, OECD and World Bank differ in important respects. The IMF is primarily concerned to provide statistics on trade, payments and financial matters (as in its *Direction of Trade Statistics* and *Balance of Payments Statistics*). It provides detailed debt-related data mainly on debts owed to the IMF in the form of IMF credits. The Basle-based Bank for International Settlements was set up in 1930; in effect, it is the central bankers' bank. The BIS 'periodically reports liabilities and assets of borrowing countries to commercial banks in the Group of Ten countries, some other industrial countries and to certain of their foreign affiliates in major offshore banking centres' (Nunnenkamp 1986, 25). Although short-term debt is included in BIS statistics, the BIS does not consider debts owed to creditors other than the commercial banks. This is an obvious limitation on BIS data as a guide to the total debt stocks and flows which affect developing countries, and particularly those whose debts are due mainly to the non-banking sector. The OECD offers a more comprehensive debt reporting system than does the BIS. Figures are collected on public and private debt, whether or not it is guaranteed. Unlike in the BIS data, short-term debt is excluded.

The fourth reporting system is that run by the World Bank. The Debt Reporting System (DRS) of the World Bank was set up in 1951 and it remains the 'sole repository for statistics on the external debt of developing countries on a loan-by-loan basis' (World Bank 1990a, ix). The data are collected and maintained by staff of the Debt and International Finance Division of the International Economics Department (IECDI). This division is charged with collecting data reported to the DRS by countries which are members of it (111 in 1990), with aiding countries in the preparation of this data, with providing estimates of debt stocks and flows for countries outside the DRS, with providing consistent aggregations and transformations of data received, and with cross-checking and supplementing World Bank data with debt statistics collected by the BIS, OECD and IMF. A common criticism of the World Bank's DRS is that it records only long-term debt and that, 'for many countries only public and publicly-guaranteed debt is presented' (Edwards 1988, 6; see also Nunnenkamp 1986, 27). This criticism is now out-dated. The DRS of the World Bank was improved and made more extensive in the 1980s. Since the mid-1980s data on short-term debts have been consistently displayed (and displayed for the late 1970s and early 1980s as well). Data are also provided for private non-guaranteed debts. Finally, steps have been taken to express debt data in a common currency (the US dollar), notwithstanding the difficulties that this entails in terms of setting meaningful average and year-end exchange rates (see section 3). Although other difficulties beset the DRS, the World's Bank's *World Debt Tables* offer the most accessible, reliable and comprehensive set of statistics on the external debts of developing countries. It is for this reason that most of the statistics presented in this book are taken from World Bank publications.

3 Definitions

Figures on the debt crisis are discrepant not only because of the varied systems of data collection from which they derive; they also differ according to particular definitions of what is being measured. 'Debt' itself is a case in point. To most ears, the word debt has a negative ring to it; it is something to be avoided. Further reflection suggests that this is too simple. The ability of a household, or a firm, or even a country to go into debt is rather a sign of creditworthiness. It is when we do not qualify for credit that we need to be worried. (This assumes that the transaction is on a voluntary basis; such a statement would not extend to debt bondage.) Put another way, if it is money that 'permits the separation of sales and purchases in space and time' (Harvey 1982,

245), it is credit that makes possible large-scale economic activities and exchange. Without credit – without the capacity to go into debt – it would be difficult to buy a house (take out a mortgage), build the Channel Tunnel (by borrowing on the financial markets) or 'develop' (with the help of foreign aid, bonds, bank loans, etc.).

It follows from this that the level of debt outstanding for a given borrower is not always a telling statistic. More debt is not always the same thing as worse debt. In March 1990 the United States owed the rest of the world 1061 billion US dollars, or about as much as the developing country debtors owed to their creditors collectively. But the USA also had claims on the rest of the world which totalled $924 billion. These are its overseas assets, which have been built up over many years. The net external debt position of the United States is then $137 billion (which is still the largest debt of any nation). More to the point, the USA has no difficulty in servicing the repayments which continually fall due on this outstanding stock of debt.

This is a critical point of distinction. When we talk about the 'debt crisis' we are not talking mainly about the huge stocks of debt which are owed by a given country to banks and other creditors without its borders. No one seriously expects the principal of Brazil's debt to be repaid in full, and no one wants the USA quickly to repay all of its net external debt. If the USA sought to pay off its external debt by running a trade surplus there would be large-scale and damaging repercussions for existing patterns of industrial trade. What matters is the commitment which a country makes to service its debt repayments. The financial system works according to a series of flow mechanisms, with banks seeking to earn interest on their assets (or loans made to creditworthy customers). The system works as long as bank loans are properly serviced and money is kept in a continuous rotation. If the rotation is threatened or stopped, the banks and their clients can run into difficulties. In most industrial countries banking regulations are such that a loan is declared to be non-performing if repayments on it are not made within a given number of days from a due date. This has implications for the bank's balance sheet and for its own creditworthiness. The prospect of non-performing loans is a fear which most banks lived with in the 1980s developing countries' debt crisis. This prospect is less evident in the case of other large debtor nations –

the USA, Australia, France – precisely because debt servicing is not expected to be a problem.

The distinction between the stock and flow components of debt is central to the way in which debt statistics are collated. The DRS of the World Bank presents statistics both on 'debt stock and its components' and on 'aggregate net resource flows (long-term) and net transfers to developing countries'. We can consider them in turn.

The total external debt stock (EDT) of a country 'consists of public and publicly guaranteed long-term debt, private nonguaranteed long-term debt (whether reported or estimated by staff of the Debt and International Finance Division), the use of International Monetary Fund credit, and estimated short-term debt' (World Bank 1990a, xii; see figure 2.1). In recent years, interest on arrears

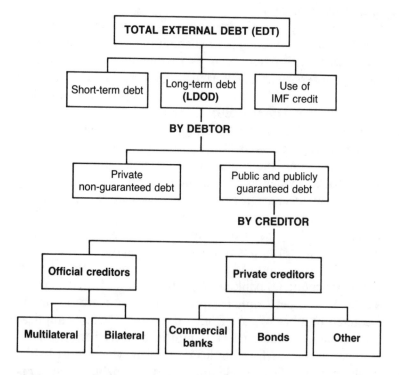

Figure 2.1 Debt stock and its components
Source: World Bank (1991a)

on long-term debt has been included and is shown as a separate entry.

Use of IMF credit is calculated on a year-end basis and refers to all repurchase obligations to the IMF. These repurchases might have been made under the main credit tranches of the IMF, and/or by means of its various special facilities (such as the buffer stock, the compensatory financing facility and the oil facilities). It also includes transactions under the terms of the Structural Adjustment Facility and the Enhanced Structural Adjustment Facility. Many of these facilities were set up in the wake of the developing countries' debt crisis, and some of their workings will be explained later on in this chapter.

Short-term external debt is defined as debt which has an original maturity of one year or less. The World Bank receives information on short-term debt from some individual reporting countries, but it acknowledges that such debt stocks are difficult to measure and to monitor. Many debtors do not report short-term debts. World Bank staff are then forced to provide estimates based on data which are available from creditors and reported to the BIS. Although the reporting of short-term debt is improving, it remains an obvious point of weakness in most debt reporting systems. In the early 1980s information on short-term debt was especially poor, with consequences that I shall report on in sections 5 and 6.

Finally, there is the category of long-term external debt, which remains the most important stock category in terms of totals of money disbursed. The World Bank defines long-term external debt as 'debt that has an original maturity of more than one year and that is owed to non-residents and repayable in foreign currency, goods or services' (World Bank 1990a, xiii). Long-term external debt has three components:

1 Public debt, which is an external obligation of a public debtor, including the national government, a political subdivision (or an agency of either) and autonomous public bodies.
2 Publicly guaranteed debt, which is an external obligation of a private debtor that is guaranteed for repayment by a public entity.
3 Private non-guaranteed external debt, which is an external obligation of a private debtor that is not guaranteed by a public entity. (After World Bank 1990a, xiii.)

Not surprisingly, the mix of public, publicly guaranteed and private non-guaranteed external debt in the debt stocks of particular countries varies a great deal. This can have implications for the scale and timing of repayment flows due on particular debt stocks. Debtors also vary in terms of the type of creditor to whom they are indebted. In simple terms, creditors are either official creditors or private creditors. Official creditors comprise bilateral lenders (where one country lends to another) and multilateral lenders (as when debts are incurred to the World Bank, the European Community or the Inter-American Development Bank). Private creditors comprise the commercial banks (in debt crisis circles reference is often made to the US money-centre banks; Citibank, Chase Manhattan, Bank of America and so on), the bond markets and bond holders, and a catch-all category of 'others' (including credits from manufacturers and exporters). The question of which creditor a repayment is first owed to is not unimportant. In debt reschedulings the first claim on the resources of a debtor country is exercised by the IMF. Other creditors enter into negotiations to establish repayment arrangements and an order of priority. (Generally speaking, the debt crisis in Latin America is linked to a private banking crisis. It is the debt crisis we all know about, and the main forum for negotiations in regard to it is the so-called London Club. The debt crisis in sub-Saharan Africa is linked to a crisis of development and an inability to finance returns on debt to official creditors. The co-ordinating institution in this case is the Paris Club.)

Let us now turn to the net resource flows and transfers which correspond to given stocks of external debt. Keep an eye on the first and second columns of figure 2.2. Reading downwards we see that loan disbursements (or drawings on loan commitments in a given year) minus loan amortizations (or repayments of principal in a given year) equal an annual net resource flow on debt. Even in the mid-1980s, this flow was positive in Latin America and the Caribbean, as it was in other indebted regions. New disbursements of long-term debt and IMF purchases more than covered principal repayments on long-term debt and IMF repurchases. But this is not the end of the story. The debt is serviced in a second way. Loans are made at a rate of interest. In effect, the interest rate is a measure of the price of money, and from the point of view of the lender it

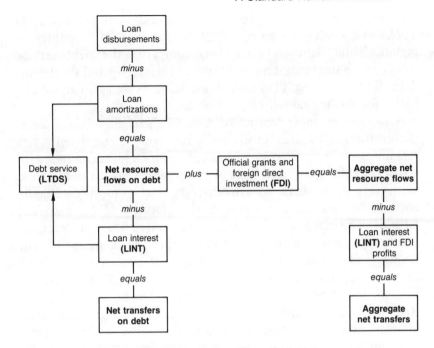

Figure 2.2   Aggregate net resource flows (long-term) and net transfers to developing countries
Includes only loans with an original maturity of more than one year.
Excludes IMF transactions.
Source: World Bank (1991a)

must take into account likely credit risks and inflation trends as they affect the borrower. Debt service then comprises loan amortizations plus loan interest (LINT). More significantly, the net resource flow on debt minus LINT equals the net transfer on debt. This is an important statistic on debt from the point of view of the indebted nation. In the highly indebted countries, net transfers on debt were negative for most of the 1980s. More money was leaving a given debtor country than was entering it.

Two further points are worth noting. First, the net resource flow on debt of a country is often added to by official grants and foreign direct investment (FDI). The aggregate net transfer to a developing country then consists of an aggregate net resource flow (net resource flow on debt, plus grants and FDI) minus LINT and FDI profits remitted to the host nation. This statistic will not greatly concern us in this book. Second, the World Bank's *World Debt Tables* make reference to the 'total debt service' of a country or region. Total debt service (TDS) comprises the debt service payments on total long-term debt, use of IMF credit and short-term debt. By subtracting TDS from disbursements we again arrive at a figure for the annual net transfer on debt.

A worked example will make some of these distinctions clear. Before turning to such an example, however, let us detail a few remaining debt indicators which are commonly encountered in statistical collections, and in books and papers on the debt crisis. With respect to loan repayments, the terms most commonly encountered are the grace period, the maturity of the loan and LIBOR. The grace period refers to the period of time which elapses before repayments of the principal, or capital sum, are made. The maturity of a loan refers to the period of time over which all repayments are expected to be made (or amortized: from the Old French, to extinguish). LIBOR is the London Interbank Offered Rate. Just as the bank rate or prime rate offers a guide to likely interest rates for mortgage-holders in the UK and USA respectively, so interest rates on external debts are quoted in terms of the interest rates which one bank charges another (three-month LIBOR, six-month LIBOR etc.), and certain spreads (or risk premiums) above it: say six-month dollar LIBOR plus 9/16th of one per cent. Again, the significance of these terms will become apparent in section 4.

A second set of debt indicators are referred to as the principal ratios. Precisely because figures on the total external debt of a nation do not always tell us very much, scholars, planners and politicians have turned to simple ratio measures to provide an indication of the severity of a country's debt problems. None of these measures is infallible and it is unwise to look to any one or two of them in isolation. Nevertheless, papers and books (including this one) will commonly make reference to debt/export ratios (which can take the form of total external debt divided by a country's exports of goods and services (EDT/XGS), or total interest payments divided by the exports of goods and services (INT/XGS). Other ratios commonly referred to include total debt service divided by exports of goods and services (TDS/XGS); total external debt *or* total interest payments divided by gross national product

(EDT/GNP, INT/GNP), international reserves divided by total external debt (RES/EDT) and short-term debt divided by total external debt (short-term/EDT). Similar calculations can be done on long-term debt only, as opposed to total external debt.

Finally, an important technical note of caution is worth entering. Most countries hold debt stocks in different currencies (mainly US dollars, but also sterling, yen, Deutschmarks, French francs and so on). Data on debt are normally reported to the World Bank in this differentiated form. The World Bank then undertakes to convert these debt statistics into a common currency (the US dollar). But this is far from straightforward. Currencies do not fluctuate against one another on a predictable year-end basis only; under a managed float they are moving up and down all the time (and in different directions against different currencies). The question of when to set a currency conversion rate can then be a critical one. The World Bank chooses to convert stock figures using end-period exchange rates derived from the *International Financial Statistics* of the International Monetary Fund. Its flow figures are converted at annual average exchange rates. This creates some unavoidable discrepancies (a point taken up by Lindert: for details see Lindert 1989, 266–70). As the World Bank puts it: 'Because flow data are converted at annual average exchange rates and stock data at year-end exchange rates, year-to-year changes in debt outstanding and disbursed are sometimes not equal to net flows. . . .Discrepancies are particularly significant when exchange rates have moved sharply in the course of the year; cancellations and the rescheduling of other liabilities into long-term debt also contribute to the differences' (World Bank 1990a, ix).

In case all this seems arcane, consider its possible relevance to the 'worked example' which closes this section (table 2.1). The data provided here are for Latin America and the Caribbean, 1970–89, and they comprise the first of four pages of data commonly presented for each country and region of the world in the annual *World Debt Tables* of the World Bank. Most of these data should now make sense. We can see, for example, that the total external debt (EDT) of Latin America and the Caribbean in 1989 is $422.188 billion (see Summary debt data). This is roughly five billion dollars less than the EDT figure for 1988, *but* most of this shortfall is the result of a depreciation in the dollar value of the

Table 2.1 Making sense of statistics on debt: Latin America and the Caribbean, 1970–1989

	1970	1980	1983	1984	1985	1986	1987	1988	1989
1 SUMMARY DEBT DATA									
Total debt stocks (EDT)	–	242535	360999	377531	389974	409708	445122	427597	422188
Long-term debt (LDOD)	27728	172788	290857	315552	330941	359396	384745	363161	348070
Public and publicly guaranteed	15855	129689	221803	249850	275414	309607	341547	332524	324558
Private non-guaranteed	11873	43099	69054	65703	55527	49789	43198	30638	23512
Use of IMF credit	128	1410	8868	11559	14548	16378	18178	16378	15748
Short-term debt	–	68337	61274	50420	44485	33933	42199	48058	58369
Interest arrears on LDOD	–	8	1198	3108	2463	3285	8393	8944	16722
Total debt flows									
Disbursements	6503	44515	37557	32570	21842	22559	22301	24277	19836
Long-term debt	6378	44100	31242	28931	19950	20451	19703	22175	16975
IMF purchases	124	415	6315	3639	1892	2108	2597	2101	2861
Principal repayments	3735	21722	15477	16398	12930	17599	17661	21317	19486
Long-term debt	3437	21235	15277	16172	12516	15667	14302	18353	16384
IMF repurchases	298	487	200	227	414	1932	3360	2964	3102
Net flows on debt	2767	22793	22080	16172	3830	1517	8115	8309	2945
of which short-term debt	–	–	–	–	–5081	–3442	3476	5350	2595
Interest payments (INT)	–	24276	34760	35229	34766	29903	28276	33284	25795
Long-term debt	1395	17292	25870	28927	28691	25266	24861	28816	20779
IMF charges	0	95	345	708	966	1128	1083	1067	1234
Short-term debt	–	6890	8545	5594	5109	3509	2332	3401	3782
Net transfers on debt	–	–1483	–12680	–19057	–30936	–28386	–20161	–24975	–22850
Total debt service (TDS)	–	45998	50237	51627	47696	47502	45937	54601	45281
Long-term debt	4832	38526	41147	45098	41207	40933	39163	47168	37163
IMF repurchases and charges	298	582	544	935	1380	3061	4442	4031	4336
Short-term debt (interest only)	–	6890	8545	5594	5109	3509	2332	3401	3782
2 AGGREGATE NET RESOURCE FLOWS AND NET TRANSFERS (LONG-TERM)									
Net Resource Flows	4159	29286	20263	16993	12942	9572	12572	13441	8873
Net flow long-term debt (ex. IMF)	2941	22865	15965	12759	7434	4784	5401	3822	591
Grants (excluding tech. assistance)	130	338	762	985	1245	1279	1514	1650	1691
Direct foreign investment (net)	1087	6083	3536	3249	4262	3509	5657	7968	6591
Net transfers	778	7164	–9368	–15620	–20014	–20494	–17170	–21251	–17905

3 MAJOR ECONOMIC AGGREGATES

Gross national product (GNP)	151623	695909	616108	635015	641436	660633	693454	798372	904563
Exports of goods & services (XGS)	18540	123078	116773	129433	123627	107802	120603	135362	148465
Imports of goods & services (MGS)	21868	154245	126346	131757	128287	126470	132799	147666	158333
International reserves (RES)	5427	57059	39526	47810	49808	43564	50718	42737	43923
Current account balance	−3139	−30475	−8396	−660	−1998	−16574	−9420	−8266	−5786

4 DEBT INDICATORS

EDT/XGS (%)	—	197.1	309.1	291.7	315.4	380.1	369.1	315.9	284.4
EDT/GNP(%)	—	34.9	58.6	59.5	60.8	62.0	64.2	53.6	46.7
TDS/XGS(%)	—	37.4	43.0	39.9	38.6	44.1	38.1	40.3	30.5
INT/XGS(%)	—	19.7	29.8	27.2	28.1	27.7	23.4	24.6	17.4
INT/GNP(%)	—	3.5	5.6	5.5	5.4	4.5	4.1	4.2	2.9
RES/EDT(%)	—	23.5	10.9	12.7	12.8	10.6	11.4	10.0	10.4
RES/MGS(months)	3.0	4.4	3.8	4.4	4.7	4.1	4.6	3.5	3.3
Short-term/EDT(%)	—	28.2	17.0	13.4	11.4	8.3	9.5	11.2	13.8
Concessional/EDT(%)	—	4.4	3.5	3.6	4.1	4.8	4.8	5.0	5.2
Multilateral/EDT(%)	—	5.8	6.0	6.2	7.9	9.9	11.5	11.8	12.2

5 LONG-TERM DEBT

Debt outstanding (LDOD)	27728	172788	290857	315552	330941	359396	384745	363161	348070
Public and publicly guaranteed	15855	129689	221803	249850	275414	309607	341547	332524	324558
Official creditors	8228	30614	45637	50356	63243	79259	98279	99414	103228
Multilateral	3001	14062	21730	23381	30788	40613	51135	50410	51698
Concessional	2032	3232	4100	4452	4809	5167	5472	5603	5805
IDA	112	427	552	580	605	649	765	873	955
Non-concessional	969	10831	17630	18929	25979	35445	45663	44807	45893
IBRD	2120	7740	11840	12010	16789	23519	30790	29577	29625
Bilateral	5226	16551	23907	26974	32454	38647	47143	49004	51530
Concessional	3971	7527	8695	9012	11042	14537	15735	15668	16355
Private creditors	7628	99075	176166	199494	212171	230348	243268	233109	221330
Bonds	1221	9612	16169	15592	17786	17509	16869	17953	22160
Commercial banks	3085	76782	140119	165003	173908	189466	201693	191211	175860
Other private	3321	12681	19879	18900	20477	23373	24706	23944	23310
Private non-guaranteed	11873	43099	69054	65703	55527	49789	43198	30638	23512
Memo: total commercial banks	14958	119881	209173	230705	229436	239255	244891	221849	199373

Values are in US$ millions, unless otherwise indicated.
Source: World Bank (1990a)

non-dollar debt stock. Within this total debt stock, some $348.070 billion of debt outstanding consists of long-term debt, mainly public and publicly guaranteed. Compared to 1988, the amount of private non-guaranteed debt has declined sharply. Section 5 will suggest why this might be the case and what its consequences might be. Short-term debt, meanwhile, has grown rapidly to $58.369 billion.

In terms of total debt flows, $19.836 billion worth of disbursements entered the region in 1989. Principal repayments totalled $19.486 billion, which left a net flow on debt of $350 million, plus $2.595 billion of short-term debt (a total of $2.945 billion). Interest payments were a further $25.795 billion, which means that the net transfer remained negative at −$22.850 billion. This figure is still less than the total international reserves of the region (item 3, RES), but these reserves (at $43.923 billion) are none the less low and offer little in the way of a cushion against unforeseen events and exogenous shocks. The ratio of total external debt to exports of goods and services (EDT/XGS) remains high at 284.4 per cent (although declining), while INT/XGS is down significantly to 17.4 per cent. Finally, in terms of a more detailed accounting of long-term debt outstanding by creditors, item 5 records that $175.860 billion is owed to private commercial banks. This is over 50 per cent of the total long-term public and publicly guaranteed debt of the region. Other regions (notably South Asia and sub-Saharan Africa) would tell a different story, as we shall see.

THE DEBT CRISIS TAKES SHAPE

4   Debt Trends: 1945–1982

It is a common expectation in economic theory that capital will flow from more developed regions to less developed regions. This was the case in the nineteenth century when British capital helped to develop the space-economies of North America, South Africa and Australasia (if not the non-white colonies from which resources were often removed in net terms). It was the case again in the 1920s when US and European capital flowed into Latin America (Stallings 1987), and it was especially the case between 1950 and 1980 when increasingly large sums of money and FDI made their way to the developing world (and to the newly industrializing countries (NICs) and middle income countries (MICs) in particular).

It is equally the case that this long history of capital flows has been associated with a long history of periodic default. In 1839 the states of Mississippi and Louisiana 'defaulted on their debts, mostly to Britain . . . and their subsequent refusal to reach a settlement with their creditors led to the formation of the Council of Foreign Bondholders in 1868' (Congdon 1988, 110). In the 1870s there was a chequered history of default, with a number of Latin American countries failing to service their loans, along with similar tales of non-payment in Turkey and Egypt. Sixty years later this was followed by a wave of defaults by 17 Latin American countries, as commodity prices tumbled in the wake of the Great Crash of 1929. 'Only Argentina honoured its debts in full and on time during that decade [the 1930s] and the 1940s' (Congdon 1988, 110).

Needless to say, these instances of default are only the tip of an iceberg. If they do not tell the full story, they at least lend perspective to the remarks of those who believe that the 1980s debt crisis is a new phenomenon, or that defaulters are to be found only in developing countries. We can be equally sure that the difficulties of the 1980s will not discourage less developed countries (LDCs) from seeking further net transfers of funds from capital markets and corporations in the developed world (and that some will get into difficulties as a consequence).

There is, none the less, an important set of distinctions to be drawn between the types of debt instruments and creditor–debtor relationships which were common in the 1920s and 1930s and those common in the 1950s and 1960s and the 1970s and 1980s. In the 1920s most lending to Latin America and elsewhere was through the bond markets, with country debt being held by thousands of individual bond-holders based in many different countries. When Latin American countries defaulted on their debts in the 1930s the loss was borne by the bondholders. The effects of this debt crisis were correspondingly fragmented; the defaults did not threaten, disproportionately, ten or twenty large lending institutions occupying particular positions of power within the global financial system. The 1930s were not quite the 1980s.[1]

After the Second World War the main channels of lending to an emerging Third World were reshaped. In the wake of the collapse of the international bond markets, and following the apparent success of the Marshall Plan for European Recovery, the USA and some other Western nations began to make funds available to some developing countries in the form of foreign aid (or official development assistance (ODA)). Some of this aid took the form of grants, or gifts, but most of it took the form of loans with a substantial element of concessionality. In the mid-1950s the Soviet Union set up its own aid programme to compete with that of the USA, in the process confirming that aid was bound up with geopolitics. Both superpowers maintained that aid was for economic development and that aid would fill certain 'gaps' which were evident in the economies of developing countries. Sometimes these gaps took the form of foreign exchange and technology gaps; more often they took the form of a savings gap. In the 1950s and 1960s, most students of development assumed that economic growth would be secured on the basis of public and private sector industrialization programmes funded by massive and directed flows of local and external savings. This was the logic of the Harrod–Domar model, of Lewis's Two-Sector model and of Rostow's work on the stages of economic growth. Official development assistance was supposed to aid the take-off into self-sustaining growth which Rostow had identified as the pivotal stage in the development process (Rostow 1960). Foreign aid would plug the gap.

The scale of ODA increased substantially between 1950 and 1970 (although not always as a percentage of a donor country's GNP, or as fast as other forms of resource transfer). In the 1950s, the USA made available more than $20 billion of ODA. This amounted to two-thirds of the aid budget of the West and comprised about one-third of the total transfer of resources from the First World to the Third World. The other main source of resource transfer in the 1950s was FDI, or the internationalization of capital courtesy of the transnational corporations (TNCs). In the 1960s, total disbursements of ODA were much increased as other nations and institutions joined the superpowers in their various aid-giving activities and ambitions. By 1970, over 30 per cent of ODA was given by multilateral donors like the World Bank and the regional development banks.

The 1970s saw post-war patterns of resource transfer change for a second time. Although flows of ODA continued to grow (and not least from McNamara's World Bank and to sub-Saharan Africa), ODA declined rapidly in the 1970s as a percentage of total resource transfers. The 1970s will be remembered as the decade of the international bank loan, or the transfer of funds from a syndicate of Northern banks to Latin American (and some other) countries and to Southern public and private-sector organizations.

There is some truth to this remembrance, although the full picture is more complex. Figure 2.3 confirms that a switch from official to private sources of funds did characterize the 1970s in dollar terms and for a particular group of countries. This much is especially apparent in the case of Latin America and the Caribbean, although even here it is a trend more noted in some countries (Brazil, Mexico) than in others (Bolivia, Colombia). In 1970, 48.1 per cent of Latin America's long-term public and publicly guaranteed debt came from private creditors, with 19.5 per cent on loan from commercial banks. When private non-guaranteed debt is included, the proportion of debt outstanding to the commercial banks rises sharply to 53.9 per cent. This tells us something about the structure of bank lending. In 1982 the corresponding figures are 77.2, 61 and 71.2 per cent.

Other regions have different tales to tell. In North Africa and the Middle East, as in South Asia and sub-Saharan Africa, by far the major part of the debt stock outstanding in 1970 was owed to official creditors (and mainly to bilateral donors). In Europe and the Mediterranean, and East Asia and the Pacific, an early involvement with the commercial banks is once more apparent. By 1982 the story is different again. In East Asia and the Pacific the commercial banks hold close to one-half of the debt stock outstanding in 1982, as they do also in Europe and the Mediterranean. Private banks are especially active in South Korea, Malaysia, the Philippines, Indonesia, Greece, Hungary and Israel. In North Africa and the Middle East, and in South Asia, by contrast, the switch from official to private sources of credit is not so advanced. Local debt stocks have increased, but they are owed mainly to the same bilateral donors to whom these countries were indebted in 1970 (in North Africa and the Middle East for obvious geopolitical reasons), or, increasingly, to the multilateral organizations (South

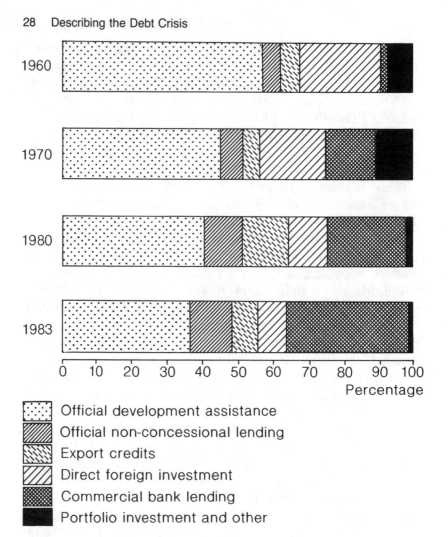

Figure 2.3 Composition of net flows to developing countries, 1960, 1970, 1980 and 1983
Source: World Bank (1985b, 20)

Asia). Finally, there is sub-Saharan Africa. In this region bilateral and multilateral ODA flows were greatly increased in the 1970s, a fact that was to haunt parts of the region in the 1980s when a series of development failures caused problems for the servicing of outstanding debt.

These regional tales caution us against any crude equation between private bank lending and national indebtedness. Even so, the two are closely linked in dollar terms and in the popular imagination, and there was an explosion in commercial bank lending to developing countries in the 1970s and early 1980s. Between 1970 and 1982 it is estimated that America's nine money-centre banks lent more than $80 billion to the non-oil developing countries. The question which must be asked is 'why?'

According to a simple version of the standard narrative account of the debt crisis the answer lies with OPEC, but this will not do. The OPEC oil price crisis of 1973 did have serious repercussions for some non-oil developing countries, and a good proportion of the wealth which flowed into OPEC coffers was recycled through the world's xenomarkets (or Euromarkets as they are more conventionally described). It is difficult, however, to sustain an argument such that the OPEC deposits made possible or necessary a huge flow of commercial bank loans to certain MICs and NICs. This much is apparent from figure 2.4, which charts the growth of syndicated Eurocurrency lending to developing countries between 1972 and 1984. The years 1973–6 do not stand out as a clear divide in this period of international financial flows. It is rather the case that OPEC funds added to a growing pool of funds which lay beyond the regulatory authorities of particular national governments.

The Eurobanking phenomenon is central to the story of modern finance and debt, and a little detail on it is unavoidable. Simply put, money is created and put into circulation by banks and other financial organizations making loans on the basis of deposits, and as a multiple of the monetary bases provided by a given central bank (e.g. the Bank of England, the Federal Reserve Bank). Insofar as there is a close link between a central bank and the commercial banks within its jurisdiction, the money supply can be fairly tightly controlled. A traditional mechanism of control in the UK financial system was the cash reserve requirement. This requirement mandated commercial (or joint-stock) banks like Barclays and Lloyds to hold a certain percentage of their deposit liabilities as cash reserve assets in their own vaults and on deposit with the Bank of England. The reserve requirement ensured that there were finite limits upon the ability of a commercial bank to make deposits (or lend money).

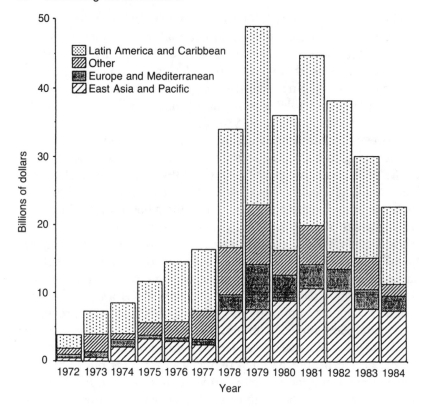

Figure 2.4   Syndicated Eurocurrency lending to developing
countries, by region, 1972–1984
'Other' includes sub-Saharan Africa, China, India, the Middle East,
North Africa and South Asia.
Source: Adapted from World Bank (1985b, 118)

In recent years this principle has been undermined and the
banking system has become much less regulated. The key to this
change lies in the internationalization of the banking industry. Put
simply, again, xeno- or Eurobanking began when the USSR and
some East European countries started to hold their dollar earnings
in those banks in Europe willing to accept deposits denominated in
the US dollar. At the time of the Cold War it made sense for the
Soviet bloc to keep its money away from the US banking system
and its government regulators (Wachtel 1986). The Euromarket
then expanded in the 1960s when the Johnson government at-

tempted to impose restrictions on the growth of foreign lending by US banks, and took steps to control domestic interest rates. The intention of these changes was to keep US capital at home, but in practice they encouraged US companies to finance their overseas operations from US and non-US banks operating beyond the confines of US banking legislation. The most rapid growth of the Euromarket, however, was in the 1970s and early 1980s. This might lead us to suppose that the market was financed mainly by recycled petrodollars, but this would be too simple. It has been estimated that the dollars deposited by OPEC countries in the Euromarkets amounted to less than 20 per cent of the total deposits taken in the 1970s. A far more important source of dollars was the balance of payments deficits run by the USA throughout most of this period. The inflation which was induced by these US deficits was also responsible in part for the growing pool of dollars available for on-lending by the commercial banks pre- and post-1973.

There are a number of reasons why certain banks became heavily involved in syndicated loans to Latin American and other MICs in the 1970s and early 1980s. There is, first, the question of a breakdown in the preceding 'aid regime' (as Robert Wood, 1986, describes it). In August 1971, President Nixon unhinged one of the main pillars of the Bretton Woods system when he broke a commitment on the part of the USA to exchange gold for dollars at a fixed rate. This rupture, together with a move to floating exchange rates in 1973, was part of a more general deregulation of international economic affairs. The old aid regime suffered within this new framework. As ODA began to multilaterialize, and to remove itself from direct US control, so the USA began to take a more jaundiced view of development through aid-led import-substitution industrialization. The new preference was for a more outward-looking development which would be part-financed by private funds from abroad. The commercial banks, according to this view, could more efficiently fulfil the role played previously by the aid organizations. Following the OPEC price shocks, the US banks were expected to do their bit for development, for international economic stability and for America's geopolitical ambitions in Latin America (Frieden 1987a).

The fact that the banks might have been pushed to play such a role does not mean that most banks were unwilling to act as brokers between the developed and developing worlds. The US money-centre banks had their own reasons for extending their operations in some middle-income countries (Makin 1984). One reason concerned the recession, which was biting hard in many OECD countries following the events of 1973. Insofar as most money-centre banks were dependent upon corporate clients at this time, the recession meant that demand for new loans was declining sharply. The banks were taking in deposits but had few obvious clients for their loans. This mismatch was compounded by the (re-) emergence of commercial paper markets in the early to mid-1970s. The world's largest corporations were now minded to raise money by issuing their own bonds and securities, rather than by seeking loans at higher rates of interest from the commercial banks.

In this context the Third World began to seem like an attractive proposition to many banks, with Latin America in particular seeming to offer a source of profits which was not available elsewhere (and for which the US and European banks began to engage in a fierce struggle with their Japanese rivals). Above all, these profits looked secure. Walter Wriston, the Chairman of Citibank at the time, is famous for his dictum that countries never go bankrupt (Wriston 1986). But Wriston's philosophy was hardly a singular one. Other banks followed the lead of Citibank for apparently sober reasons: most loans would be made at floating interest rates (which left the borrower with most of the risks); most loans would be made at a sizeable spread above LIBOR (an indication that risks were involved); and most loans would be made by a syndicate of banks, fronted for a fee by a lead bank such as Citibank or Chase Manhattan. Finally, there was some truth in Wriston's aphorism: most of the loans made in the 1970s and early 1980s were made to governments or to public sector agencies and corporations. Although one might have expected US and other banks to have been sceptical of state-capitalist development programmes, fifteen or twenty years ago such scepticism was not common. An ideology of public sector developmentalism was preferred by the banks because it seemed to offer the requisite guarantees for the loans which were being made. Non-guaranteed loans to the private sector were much more of a risk.

Whether or not these were sound reasons for lending on the scale observed for the period 1977–82 is a moot point (and one that we will return to in chapters 3–5). In hindsight, it is apparent that the risks endemic in bank lending were expanded by three particular sets of factors.

There is, first, the question of the geography of commercial bank lending. Between 1973 and 1982, 'the total gross outstanding debt of Latin America and the Caribbean grew at a compounded annual rate of 25 per cent . . .or at almost twice the rate of growth of export earnings and about four times that of the GNP' (Kuczynski 1988, 36). Of this debt, fully three-quarters of the increase came from the commercial banks, and of this $70 billion, some $35 billion took the form of transactions between just three countries (Brazil, Mexico and Venezuela) and nine US money-centre banks (see also table 2.2 for 1984 data). Commercial bank loans were also sizeable to South Korea, to Hungary and to Nigeria and the Côte D'Ivoire, although in each of these countries the dominance of US banks was not so apparent. German banks were closely involved with some East European debtors, while a mixture of private and official funds were channelled into Turkey, Egypt, Indonesia and the Philippines. South Asia and sub-Saharan Africa were not deemed sufficiently creditworthy to attract large-scale commercial bank loans, although in each case loans were available from official creditors.

A second set of factors has been referred to as greed, naivety, mismanagement and inexperience. In less compelling language, we might note that the mechanics of commercial bank lending overseas in the 1970s were not especially risk-averse. At this time it was not uncommon for syndicated bank loans to be made on a commission basis, with the lead bank taking as its fee one-eighth of one per cent of the value of the loan disbursed ($1.25 million on a $1 billion loan). It was also not uncommon for loans to be made by young men apparently unschooled in the more principled activities of safe lending. S. C. Gwynne offers a colourful account of the life of a young banker in the 1970s when he writes of his 'adventures in the loan trade'. Says Gwynne:

> I am far from alone in my youth and inexperience. The world of international banking is now full of aggressive, bright, but hope-lessly inexperienced lenders in their mid-twenties. They travel the world like itinerant brushmen, filling loan quotas, peddling financial

Table 2.2 The weight of Latin American debt in the nine leading US money-centre banks

	Mexico	Brazil	Venezuela	Argentina	Chile	Five countries' total	Five countries' loan as % of shareholders' equity
Bank America	2.7	2.5	1.5	0.5	0.3	7.5	145.1
Citibank	2.9	4.8	1.4	1.2	0.5	10.8	178.6
Chase Manhattan	1.6	2.7	1.2	0.8	0.5	6.8	198.3
Manufacturers Hanover	1.9	2.2	1.1	1.3	0.7	7.2	254.7
Morgan Guaranty	1.2	1.8	0.5	0.8	0.3	4.6	134.5
Continental Illinois	0.7	0.5	0.4	0.4	0.3	2.3	124.5
Chemical	1.4	1.3	0.8	0.4	0.4	4.3	179.6
Bankers' Trust	1.3	0.7	0.4	0.3	0.3	3.0	166.8
First Chicago	0.8	0.7	0.2	0.2	0.2	2.1	116.3
Total	14.5	17.2	7.5	5.9	3.5	48.6	166.5

Value of outstanding loans in US$ billions, end 1984
Source: Anatole Kalestsky, *The Costs of Default.* New York: Priority Press, 1985

wares, and living high on the hog. Their bosses are often bright but hopelessly inexperienced twenty-nine year old vice-presidents with wardrobes from Brooks Brothers, MBAs from Wharton or Stanford, and so little credit training that they would have trouble with a simple retail instalment loan. (Gwynne 1983, 23)

This story may be apocryphal, but it is not the only story of this type to emerge from the 1970s (or from the 1980s: see Wolfe (1987) on the 'Masters of the Universe'; and see Lewis (1989) on the 'Big Swinging Dicks' at Salomon Brothers). Certainly, Edwards is right to maintain that the Eurocurrency markets developed by virtue of 'competition in "fiscal laxity" with the commercial banks based in the UK, Germany, Japan, and so on, successfully pressurizing their central banks to allow them to compete internationally by *not* being obliged to carry the reserves on overseas operations which applied to their "home" operations' (Edwards 1985, 179; emphasis in the original). In short, it was the structure of the Euromarkets, with a credit multiplier tending in theory (if not in practice) to infinity, which laid the conditions for an apparently institutionalized avarice and recklessness. Moreover, these attitudes or traits were not confined to the commercial banks. The LDCs which took out commercial bank loans were usually pleased to do so. For one thing, the loans seemed to make possible the very 'development' which was expected of such countries (see chapter 3). The loans also had more obvious attractions: they were not always tied to particular projects or to particular groups of suppliers (as was often the case with foreign aid); they were contracted at rates of interest which were often below the prevailing rate of inflation in the creditor countries (in which case the real interest rate was negative); and their nominal stock value appeard to be devaluing rapidly on account of the very inflationism from which the loan climate had emerged. Finally, such loans were often critical, 'in closing the circle started by high oil prices and a weak dollar' (Kuczynski 1988, 15). The loans made economic growth and export expansion possible in some developing countries precisely when a recession was biting hard in the industrialized world. In effect, the loans substituted for (or, better, added to) ODA and the rather meagre domestic savings which were generated within the unequal class structures which beset most borrowing countries.

A third set of factors concerns the nature and terms of the financial transactions themselves. Although money was cheap in the period 1971–9, this changed markedly between 1979 and 1982 (see section 5). The structure of indebtedness also changed. In the mid-1970s, 'maturities to twelve years were in some cases possible, but for most of the decade eight years was the normal limit, with some loans of up to ten years' (Kuczynski 1988, 76). These maturities are low in comparison to the typical amortization period on ODA and some countries in the 1970s tried harder than others (Brazil more than Mexico) to keep 'maturities as long as possible' (ibid.), even at the expense of higher spreads on the interest rates which attached to a loan. Nevertheless, these loans can reasonably be described as long-term loans.

After 1979, a different debt profile begins to emerge. As money became scarce, interest rates began to rise. Meanwhile, as some banks at last began to have some doubts about the loans they had made to some developing countries, spreads became higher and final maturities were shortened. The banks now sought to lend at a premium and with the objective of having 'a higher porportion of loans for self-liquidating trade-related purposes, such as financing imports or exports' (Kuczynski 1988, 77). Those LDCs with large debts outstanding, and new oil bills to pay, had little choice but to comply with this sudden shift to short-term bank lending. The result was an explosion of short-term debt between 1978 and 1982. Table 2.3 offers an account of what happened in Latin America. In more detail, one might note the case of Mexico. Kuczynski recounts that:

> Between the end of 1979 and the end of 1981, Mexico almost doubled its external debt to banks; moreover, $17 billion of the $26 billion increase in two years was at short-term. This enormous increase in short-term debt took place at the very time that the average price of Mexico's oil exports went from about $13 per barrel to $36, fueling the largest increase in export earnings in Mexico's modern history. Meanwhile, however, public expenditures rose just as quickly, helped by the combined effect of the increase in external debt and the higher tax revenues from oil. The frantic pace of public expenditure . . .raised public sector outlays from Mex. $1.3 trillion to Mex. $2.8 trillion two years later, or from the equivalent of 41 to 47 per cent of the GDP. Foreign bank borrowing financed about 20 per cent of the increase. (Kuczynski 1988, 79)

Table 2.3   The growth of short-term bank debt, 1978–1982 (in billions of US dollars, outstanding at end of year)

	1978	1979	1980	1981	1982
Argentina, total	7.0	13.4	19.9	24.8	25.7
Of which short-term	3.5	6.9	10.4	11.6	13.9
Brazil, total	32.9	38.6	45.7	52.5	60.5
Of which short-term	9.3	11.3	16.2	18.2	21.1
Mexico, total	23.2	30.9	42.5	57.1	62.9
Of which short-term	7.4	10.7	18.8	27.8	29.9
Venezuela, total	14.0	20.8	24.3	26.2	27.5
Of which short-term	7.6	12.7	14.3	16.1	15.8
Latin America, total	97.1	129.0	162.9	196.6	214.2
Of which short-term	38.3	54.4	75.2	91.4	98.6
1 Rate of growth of total(%)	30.3	32.9	26.3	20.7	9.0
2 Net disbursements by banks	22.6	31.9	34.0	33.7	17.6
3 Interest paid to banks (est.)[a]	9.0	15.3	22.3	32.2	30.4
4 Net transfer from banks (2 less 3)	13.6	16.6	11.7	1.5	−12.8

Basic data are from Bank for International Settlements. These exclude offshore banking centres such as the Bahamas, Grand Cayman, and Panama. This omission probably leads to an underestimate of 10 to 25 per cent in outstandings and flows. Author's estimates of interest paid are based on LIBOR plus an average margin of 1.25 per cent. The data encompass short-term debt with an original maturity of up to and including one year.
[a]Interest received from banks on reserves deposited abroad has not been netted out.
Source: After Kuczynski (1988, 78)

5   The Crisis: 1982–1983

It is against this background that the debt crisis came to the world's attention with the decision of Mexico to default on its external debt in August 1982. In the light of what has been said so far, some of the reasons for this default are not hard to find.

The wider context to Mexico's default lies with the international economic situation between 1978 and 1982. In September 1978 the US dollar came under such pressure that President Carter was obliged to turn his back on the inflationary and devaluationist strategies which the USA had pursued throughout the 1970s. As a symbol of America's new commitment to 'sound money', Paul Volcker was installed in 1979 as Chairman of the Federal Reserve Board, a job which confers considerably more power and autonomy from government than does the position of Governor of the Bank of England (Greider 1987). The money supply of the USA was swiftly tightened by Volcker and certain versions of monetarist economics were advanced by the authorities to justify America's new monetary policy (see chapter 3). By June 1982 the prime rate in the USA had reached 16.5 per cent and unemployment in that country had risen to 9.3 per cent of the total labour force. Similar tales were to be told in Western Europe and especially in Mrs Thatcher's Britain. The so-called New Right was in the ascendant and a process of economic restructuring was under way.

The implications of these changes for the developing countries were two-fold (and varied as such from country to country). On the one hand, the six-month dollar LIBOR climbed from an average of 9.20 per cent in 1978 to an average of 16.63 per cent in 1981. Real interest rates were also on the rise and they were to remain at a high level until 1986 (see figure 2.5). William Cline has estimated that 'total excess [or unanticipated] interest payments on developing country debt amounted to $41 billion in 1981–82' (Cline 1984, 12). Insofar as a country's debt was denominated mainly in dollars this unanticipated expense was further compounded by the appreciation of the dollar from 1980 to 1985. Capital flight was also encouraged (chapters 3, 4 and 5). The other blade of an emerging scissors crisis was closed by a general decline in the value of commodity exports from developing countries from 1979 to 1987. Again, the nature and impact of what really was a variegated set of commodity price movements is the subject of heated debate (compare chapters 3–5). It is not unreasonable to suggest, however, that a decline in non-oil commodity prices placed some LDCs in a vulnerable position. Cline concludes that 'high interest rates and the global recession imposed large cumulative losses on the non-oil developing countries in 1981–82. In all, these countries lost approximately $141 billion in higher interest payments, lower export receipts, and higher import costs as the consequence of adverse international macroeconomic conditions' (Cline 1984, 13).

Whether default was inevitable in this context is again a matter for debate. Not all developing countries defaulted on their debts in 1982–3 (one thinks of South Korea, Malaysia, Colombia, Thailand and Zimbabwe), but not all developing countries faced the catalytic factors which were present in the case of Mexico and some other Latin American defaulters. Apart from domestic policy considerations (which will be examined later), the three most important of these factors were the emerging time-profiles of some Latin Ameri-

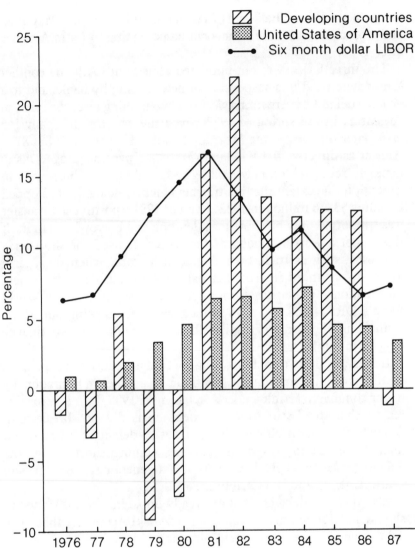

Figure 2.5 Real interest rates and LIBOR
The real interest rate is defined as the six-month dollar LIBOR deflated
by the change in the export price index for developing countries. The
US real interest rate is defined as the six-month dollar LIBOR deflated
by the US GDP deflator.
Source: World Bank (1987a)

can debt stocks, the Falklands-Malvinas conflict of April–May 1982 and a precipitate decline in new net bank lending to Latin America after 1981.

The three factors are not unrelated. The South Atlantic conflict 'immediately led to a suspension of new lending by banks and to a *de facto* default by Argentina. More than anything else, the crisis in Argentina in the spring of 1982 paved the way for the really big crisis later that year – the Mexican default' (Kuczynski 1988, 81).[2] Almost as directly, the Falklands-Malvinas war encouraged many banks to rein back on new lending to Latin America borrowers in general, to the extent that net transfers on private banking dropped to minus $3816 million in 1982. Even in 1981 a positive net transfer was supported only by short-term lending and borrowing. (In itself this is not a worry, as long as the money is used for short-term purposes, such as trade financing. It is a worry when short-term loans are advanced in a market palpably not blessed with perfect information – either on an interbank basis or through the World Bank's DRS – and where the funds are either flowing out of the country in the form of capital flight, or being used to finance long-term public deficits.)

In sum, if 'short-term borrowing . . .did not cause the debt crisis [it did enable] borrowing to continue at a time of hesitation by major commercial bank lenders' (Kuczynski 1988, 80). It is for this reason that the Latin American debt crisis did not break until August 1982 (in an official sense, and with defaults by Brazil and many other Latin American countries coming shortly after the default by Mexico). By 1984 there was a generalized debt crisis in Latin America and the Caribbean.

Elsewhere, a debt crisis was already apparent by 1981 and we should not forget this. Although Kuczynski maintains that 'The debt issue is, to a large extent although not exclusively, concentrated in Latin America' (Kuczynski 1988, 1), this judgement is at odds with an historical geography of default on mainly non-commercial debts which peaked in the early 1980s. In this regard it is not just the case that Zaire defaulted on its debts in 1975, which is commonly enough cited; it is also the case that, between 1979 and 1983, 'twenty Third World countries renegotiated their debt to bilateral official creditors. . . . Indeed, up through 1982, there were considerably more Paris Club reschedulings of debt to bilateral official creditors than there were reschedulings of debt to private creditors' (Wood 1986, 234). If the Mexican default was a prelude to a widespread debt-cum-banking crisis, 'Several years before Mexico's announcement a widespread but little-noticed official debt crisis had emerged' (ibid.). The debt crisis, which had also claimed Poland in 1981, was set to take many forms and to call forth many different proposals for its amelioration.

DEBT CRISIS MANAGEMENT

According to the standard narrative account of the debt crisis, the crisis has been managed in three distinct ways: by means of containment, austerity and adjustment (1982–5); by means of adjustment with growth (the Baker years of 1985–8); and by means of the Brady initiative and a market-menu of debt writedowns (1988/9 to the time of writing). Some will dissent from this narrative and their voices will be taken up in later chapters. Others will point to the regional limitations of this account. Those who do recognize the storyline will want to acknowledge that these three stages of debt crisis management are closely linked and in important respects are overlapping.

6   Containment, Adjustment and Austerity

The advanced capitalist countries were not slow to appreciate the significance of Mexico defaulting on its debts. When Paul Volcker spoke to the Subcommittee on International Finance and Monetary Policy of the Committee on Banking, Housing and Urban Affairs, US Senate, on 17 February 1983, he emphasized that:

> We face extraordinary pressures in the international financial system . . .This is not an abstract, esoteric problem of marginal interest to our [US] economy. Failure to address these problems will jeopardize our jobs, our exports and our financial system. Unless it is dealt with effectively, it could undermine both our own recovery and the economies of our trading partners and friends abroad. I am confident that the situation can be managed – but it won't manage itself. (Volcker 1983, 175).

Later on, Mr Volcker confirmed that 'Our concern for maintaining a well-functioning international financial system is rooted in our self-interest, not in altruism' (ibid. 176).

Volcker's remarks on altruism and its absence were intended for a particular audience; in effect, he was asking Congress to approve an increase in IMF quotas as part of a broader debt management programme. This quota increase was to be sold on the basis of self-interest rather than a splendid internationalism. Volcker's underlying confidence, by contrast, stemmed from his belief that the Mexican default had been capably contained by the USA, the banks and the IMF, and that this first attempt at debt management by containment could serve as a model for future problems and reschedulings. In this context it makes sense to return to the Mexican negotiations of mid to late 1982, in the process exploring in fine detail the debt management package which emerged then and the assumptions upon which it was based.

Kuczynski provides an excellent guide to the events of August–December 1982 in Mexico and we will follow him closely (see also Kraft, 1984). Kuczynski begins by noting that 'The Latin American debt crisis seemed to burst upon the world as a complete surprise in August 1982. Only a few top people in the Mexican Finance Ministry and a few not-so-top people in the US Treasury had been warning for several months that a Mexican default was likely' (Kuczynski 1988, 82). The story continues as follows.

1 On Thursday 12 August 1982, Mexican Finance Minister Jesus Silva Herzog telephones US Treasury Secretary Donald Regan, the Managing Director of the IMF Jacques de Larosière, and Paul Volcker to tell them that 'Mexico had almost run out of foreign exchange reserves and could no longer make its payments on its external debt' (Kyczynski 1988, 82). Herzog, his Director of Public Credit (Angel Gurria) and the Director General of the Banco de Mexico (Miguel Mancera) are called to Washington, DC.

2 Discussions over the 'Mexican weekend' (Friday 13 to Sunday 15 August 1982) make two points clear; 'first, there would be a moratorium on amortization payments of debt to commercial banks; second, in the meantime there would be an international package of emergency loans from various official sources. *An implicit but extremely important additional point was that interest*

payments would continue as scheduled' (Kuczynski 1988, 83; emphasis added).

3 Notwithstanding these two points – which constitute the beginnings of a recognizable debt management package – the full Mexican negotiations continue and remain not a little untidy and bitter. In part, this reflects a disagreement between US and Mexican negotiators on the terms of an advance US oil purchase from Mexico (which Regan hoped to obtain at a steep discount). More importantly, the negotiations are upset by the decision of the outgoing President of Mexico, Portillo, to announce the nationalization of all Mexican commercial banks and the imposition of exchange controls. Portillo is encouraged in his actions, according to Kuczynski, by Manuel Tello, a close academic advisor, and by Tello's friends and colleagues at Cambridge University, England, led by Dr Ajit Singh. Dr Singh confirms the outline of this story (conversation with author) and has several times maintained that the Mexican debt crisis need never have happened. Dr Singh prefers to blame the crisis on precipitate and discriminatory actions on the part of the US Federal Reserve and US commercial banks (see also Díaz-Alejandro, 1984 and chapter 4). Although the nationalization of the Mexican banks is not welcomed by Herzog, by incoming President de la Madrid or by the IMF, Kuczynski notes that: 'Secretly . . .many international bankers were not displeased that the debts of the Mexican banks, which had made many doubtful loans to businesses in trouble, were now the debts of the Mexican state' (Kuczynski 1988, 85).

4 Finally, the action switches to Toronto and the September 1982 joint annual meeting of the governors of the IMF and the World Bank (a jamboree which regularly attracts a retinue of more than 5000 bankers and government finance officers). The Mexican crisis is high on the unofficial agenda and dissension is again evident. Kuczynski notes that some US officials, and most obviously Beryl Sprinkel, the conservative Under-Secretary for Monetary Affairs, believe that the Mexican default is a one-off event which will be quickly patched up and forgotten. As a monetarist himself, Sprinkel is firmly opposed to the increase in IMF quotas which is agreed to at the Toronto meetings. It is only later that the position of the US Treasury changes sharply.

Specifically, it is when a crisis becomes apparent in Brazil that the stage is set for Volcker's appeal to the US Congress. Brazil seeks a moratorium on the repayment of principal to commercial banks in November 1982. It is a decision closely followed by Venezuela (in February 1983) and by Chile, Peru, Ecuador and Uruguay. In other parts of the world a not dissimilar pattern of default is emerging. By the end of 1983 rescheduling deals are in place in more than 25 countries.

The slow realization that the Mexican default might be related to a wider debt-cum-banking crisis had definite implications for the pattern of Mexico's 'rescue' (Kuczynski's word, after Kraft 1984). It suggested, in particular, that the international financial system was at risk and with it the US money-centre banks. It also helps to explain why the Latin American crises were defined as temporary liquidity crises, as opposed to more long-lasting crises of solvency. The fact that a distinction between liquidity and solvency crises is not always meaningful – can banks be solvent if debtor countries are illiquid? (see Sjaastad 1983, 317) – is not the issue here. The point is that by defining a crisis as a liquidity crisis the authorities are able to downplay a possible financial panic (including a run on the banks). It also helps to explain why and how the banks were persuaded to act in the way that they did, and why and how the banks were bound together in a Mexican rescue deal which involved the US government and the IMF.

Again, the story is important enough to bear a little detail. In 1981 the difficulties faced by Poland in servicing its external debts were met by a process of delay and by the slow setting up of a refinancing package. In the case of Mexico no such delay could be brooked. Because of the nature of banking regulations in the USA it was imperative that Mexico's ability to service the interest payments on its external debts should not be impaired. A default on these payments would threaten the solvency of some US banks. (Remember that the Bank of America Corporation and Manufacturers Hanover had lent 'more than 70 per cent of [their] common equity' to Mexico: *Economist* 16 October 1982, 23). What this meant in practice was that a package deal had to be set up which would provide emergency finance to Mexico to boost its low reserves of foreign exchange. The debts could then be serviced in part, if not in full. In the first instance the money came from '$1 billion in advance purchases by the US from Mexico of oil for its strategic petroleum reserve; $1 billion in US Department of Agriculture's Commodity Credit Corporation credits for purchases by Mexico of US staples (wheat, rice, and so on); and a package of $1.85 billion from the central banks of major European countries [including Spain], Japan and the Federal Reserve' (Kuczynski 1988, 87).

Having thus equipped Mexico, the next stage of the rescue deal involved the political management of more than 1000 banks which had at one time lent money to Mexico, and which now had to be encouraged to lend anew (if only to rollover interest payments due to themselves). Again, the US Federal Reserve took a lead in this process. On 20 August 1982, the Federal Reserve called a meeting of known bank lenders in New York City. It advised them that an advisory committee of banks should be set up to 'coordinate the effort of rescheduling . . .[bearing in mind that] the banks ranged from sophisticated banks in the major financial centres of the world to relatively small, local institutions in the hinterlands of the United States and elsewhere' (Kuczynski 1988, 88). Representatives from Citibank, the Bank of America and a European bank would play a leading role and would chair the proceedings of the Mexican bankers' advisory committee. All this happened within two weeks of the crisis erupting and eleven days before Carlos Tello took over as the new Central Bank chief in Mexico.

The third element of the Mexican rescue package involved the IMF. In the 1970s the IMF had been treated with some disdain by governments in the USA (and elsewhere). Its quota resources had dwindled from a value equivalent to 14.2 per cent of world trade in 1950 to just 3.8 per cent in 1981 (Killick 1984, 132).[3] Its powers were correspondingly weakened and mainly confined to its constitutional obligations to lend Fund resources to countries in short-term difficulties in their balance of payments. In the 1980s the position and standing of the IMF was to change dramatically, and on not a few occasions it would be made the scapegoat for austerity programmes deemed integral to a wider debt-management package, but from which governments were often keen to distance themselves. This much was already evident in the Mexican negotiations of mid to late 1982. Prior to the return to effective office of Silva

Herzog (and the inauguration of de la Madrid on 1 December 1982), the IMF was 'cleverly used . . .to soften Tello's resistance' to domestic economic and social reforms (Kuczynski 1988, 89). By 10 November 1982,

> a stabilization programme had been agreed upon with the IMF. The centrepiece was a cut in the estimated public sector deficit from the very high level of about 17 per cent of the GNP in 1981 to about 8.5 per cent in 1983, a drastic reduction. The programme thus endorsed a shock treatment for the Mexican economy, as opposed to the views of those, such as Tello in Mexico and many others elsewhere, who advocated gradualism. Clearly, economic growth was going to grind to a halt. (Kuczynski 1988, 89)

The IMF agreement made possible the final part of the Mexico deal. The main bank lenders to Mexico were called to another meeting at the Federal Reserve Bank of New York, this time on 16 November 1982. Standing before them was de Larosière, the French Managing Director of the IMF. De Larosière explained that Mexico had agreed to an austerity programme. Before the IMF Board could approve the deal, however, de Larosière insisted that 'there would have to be concrete undertakings from the bankers that $5 billion [of new loans] would be forthcoming' (Kraft 1984, 48). The banks had four weeks to give their decision, but were warned that without the new loans there would be 'no IMF standby credit and stabilization programme' (Kuczynski 1988, 90). A trap was thereby sprung. The IMF was forcing the banks to lend on an involuntary basis to Mexico or to face the consequences of not lending at all. The banks had little choice. Although some smaller banks probably were inclined to write off their Mexican loans, it was agreed that every bank lender would provide new monies to Mexico equivalent to 7 per cent of the bank's exposure. (The figure made sense: $5 billion amounted to about 7 per cent of the banks' exposure to Mexico in 1982, including short-term loans). In return for their cooperation, the banks demanded compensation in the form of substantial fees for the refinancing of Mexico's debt. They also upped the spread above LIBOR, in the process making the rescheduled loans more expensive to service. These practices continued for most of the period 1982–5.

The Mexico deal set a precedent for most of the debt reschedulings which followed in 1983 and 1984. Although the details of each rescheduling varied on a case-by-case basis, the structure of the Mexican deal was widely copied (as subsequent Mexican packages would be copied in the future: Roett 1989; Kaufman 1990). Each deal tended to take on the following shape: (a) rescheduling of the principal due to commercial banks (often over an eight-year period); (b) the maintenance of reduced trade lines and inter-bank deposits; (c) large involuntary loans; (d) significant cuts in the public sector deficits of indebted countries; (e) sharp reductions in the importation of goods and services to the indebted nation; and (f) attempts to increase the exports of goods and services from an indebted nation.[4]

The last three points require some elaboration. Public sector deficits were cut in accordance with the terms of the credit packages put together by the IMF. The thinking behind this embraced several ideas. Public sector deficits were often deemed to be a 'bad thing' in themselves, a sign of economic inefficiency and waste. More especially, it was argued that there is an intimate link between the external debt position of a country and its public sector deficit (see chapter 3). It followed that a reduction of a country's public sector deficit was a prerequisite to improving the external debt position of that country.

The argument with respect to imports and exports is more straightforward. Between 1982 and 1985 it was accepted that debts denominated in hard currencies could only be serviced in the long run on the basis of an improvement in the foreign exchange position of an indebted nation. Although steps clearly had to be taken to encourage domestic resources to be switched to the production of goods and services for export, in the short run the trading position of a country would be more easily improved by a drastic reduction in imports. This was the tack taken in Latin America and most other developing regions between 1982 and 1985.

Two more points need to be considered before we turn to an assessment of the containment years. First, there *was* a measure of concerted action on the part of the creditors, notwithstanding a commitment to a case-by-case process of debt management. Second, we have concerned ourselves thus far with Latin America. In

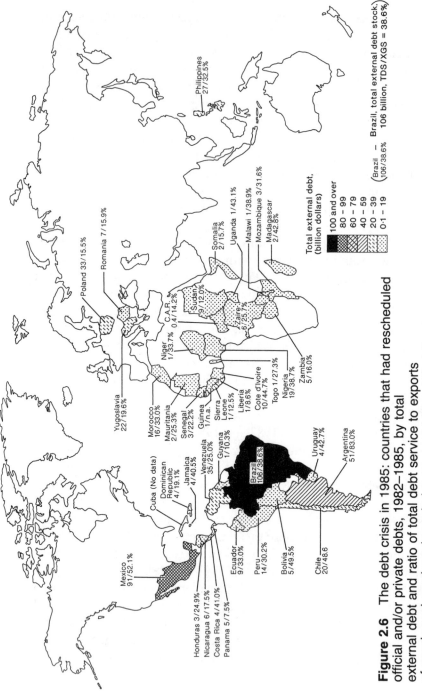

Figure 2.6 The debt crisis in 1985: countries that had rescheduled official and/or private debts, 1982–1985, by total external debt and ratio of total debt service to exports of goods and services (rounded to nearest billion US dollars)
Source: World Bank (1986a, 1990a)

part this reflects a reliance upon Kuczynski; in part it reflects the monetary value of the debts at risk in Latin America (44 per cent of total developing country external debt in 1982 and 58 per cent of outstanding private bank debt). The danger is that attention is then deflected from different types of debt crisis elsewhere: from the debt crisis in Eastern Europe, from the debt crisis affecting countries like Egypt, the Philippines and Indonesia, and from the debt and development crises affecting most of sub-Saharan Africa. In terms of the principal ratios, it is worth noting that, in 1985, the position was in some cases worse in sub-Saharan Africa that it was in Latin America and the Caribbean (see also figure 2.6).

Assessment

It is difficult to evaluate the effectiveness of the containment-austerity strategy without broaching arguments more properly left to chapters 3–5. Nevertheless, the standard narrative account of the debt crisis is sufficiently broad to entertain six particular sets of diagnoses and conclusions. These refer to: the possibility of a run on the commercial banks; the nature of the rescheduling process; the politics of negative net transfers; the performance of international trading systems; the possibility of a debt overhang and its impact on investment; and the matter of development.

The debt-bomb threat

Few articles did as much to dramatize the dangers of an imminent banking crisis as an article written by Jay Palmer that was published in *Time Magazine* in January 1983. Palmer's article was well advertised by the front cover of the magazine, which featured the Earth as a ticking bomb, with a short fuse coming out of the belly of Latin America. Palmer's thesis was that 'Never in history have so many nations owed so much money with so little promise of repayment' (Palmer 1983, 4). Palmer drew attention to the more maniacal aspects of 1970s banking practices in a way which prefigured the later work of Gwynne, Wolfe and Lewis. In this version of *The Bonfire of the Vanities*, bankers in the mid-west were lampooned by their German counterparts as 'hillbillies', while loan-market officers were described by Palmer as 'young gunslingers' (ibid. 7). It was a frightening, if often comical, vision and Palmer took care to emphasize his conclusion: that the banking systems of the Western world were in danger. The savings of ordinary depositors were at risk directly from the threat of bank failures, and indirectly from the threat of a finance-induced slump on a par with the Crash of 1929–33.

This vision did not come to pass. If we can say anything with certainty about the 1980s debt crisis it is that an international banking crisis was avoided. The major money-centre banks did not collapse as a result of the defaults in Latin America and nor will they collapse in the foreseeable future. The banks may again have to make provisions against bad debt, but the containment strategy of 1982–5 gave them time to rebuild their capital bases and diversify their portfolios. (A major catalyst to this process of rebuilding was the Reagan boom in 1980s America. By the same token, the leveraging of America in the mid-1980s is now presenting many banks with a second or third threat to their profitability – farm debt, property debt, corporate debt). By 1985 the big money-centre banks had all but survived a depression in their share prices and an assault upon their international credit ratings. In 1986 Citibank was able to issue a large number of shares, the first time that a 'large money centre bank in the US was able to raise capital in the US equity market after 1978' (Kuczynski 1988, 108).[5]

Bank lending and re-lending: the rescheduling of debt

The survival and eventual strengthening of some US banks does not mean that the position of all banks was healthy for all of the period 1982–5. Most banks were pressed by the events of the early 1980s and in the USA they did not benefit from a 'tightening of US corporate tax rules and of regulations affecting the international portfolios of banks' (Kuczynski 1988, 111). In any case, some banks fared better than others. Morgan Guaranty was able to absorb the impact of bad debts on account of its strong capitalization and its relatively high profitability. Banks forced to write down the value of assets on a large scale did not escape so easily. Kuczynski points out that there is an 'asymmetry between a modest proportion of troubled loans in total assets and a high potential impact of such loans on profits' (Kuczynski 1988, 111). A bank which is highly leveraged will find that profits are very sharply curtailed if its relatively small capital-base is written down.

This caveat entered, the argument can still be made that the containment strategy was far from bad news for all commercial banks and far from good news for all debtor countries. The reasons for this lie within the rescheduling process and they return us to the distinction between interest payments and amortization payments. Between 1982 and 1985 most attention was given to the large-scale reschedulings of the principal of Latin America's debt. The process is summarized in figure 2.7, which shows the sums involved in each rescheduling in this period and in a preceding period. Immediate maturities were first stretched over a seven to ten year period. With the introduction of Multiyear Rescheduling Arrangements (MYRAs) in 1984–5, principal repayments were set for several years (up to 1990 in the case of Mexico) and stretched to periods of twelve and thirteen years (fourteen in the case of Mexico). In due course, bank fees and commissions were also reduced and the spreads over LIBOR, which had reached 2.25 per cent in the first reschedulings, fell back to 1.50–1.25 per cent in 1984–5.

All this was progress of a sort, but it was also a case of high profile politicking which could flatter to deceive. The reschedulings of amortization tended to disguise two things. First, 'the rescheduling of amortization was not economically meaningful to the debtors' (Kuczynski 1988, 113). It would be meaningful only if there was a realistic expectation that Latin American debt was supposed to be paid back in full (or if, by implication, the US or UK national debts were expected to be 'repaid'). In a normal market this is not the case. Some lenders would expect to be repaid, while others would be happy to make new loans. By virtue of this recycling of monies, debts are continually financed. What matters is not so much the rescheduling of amortization as the rescheduling of interest payments and the associated negotiations to secure new inflows of external finance. (Rescheduling of amortization can help debtors to the extent that real debt totals may be reduced in inflation-adjusted terms.) Second, the reschedulings were expensive for the debtor nations and a sizeable source of profits for the banks. It is estimated that US money-centre banks earned $2 billion between 1982 and 1985 from fees and risk premiums on the rescheduling of loans to Latin America and the Caribbean. This is

COUNTRY	1975	1976	1977	1978	1979	1980	1981	1982	1983	1984
Argentina		970▲								23 241△
Bolivia							444▲			536△
Brazil									4 532▲	5 350△
									3 478●	
Central African Rep.							55✓		13●	
Chile	216●								3 400▲	
Costa Rica									97●	
									1 240▲	
Dominican Republic									497▲	
Ecuador									200●	4 475△
Gabon					105a●				1 835▲	590▲
Guyana					29▲			14▲		24△
Honduras										148△
India	157■	169■	110■							
Ivory Coast										153●
										306▲
Jamaica					126▲		103▲			106●
										148▲
Liberia						30●	25●	27▲	18●	17●
										71△
Madagascar							142●	103●		195▲
										120●
Malawi								24●	30●	
									59▲	
Mexico									1550b●	48 725c△
Morocco									23 625▲	530△
									1 225●	
Mozambique										200●
Nicaragua						582▲	188▲	102▲		
Niger									33●	22●
										28▲
Nigeria									1 920▲	
Pakistan							263■			
Peru				478●					450●	1 000●
					821▲				380▲	1 415△
Philippines										4 904△
										685●
Romania								234●	195d●	
								1 598▲	567▲	
Senegal							77●	84●	64●	97▲
Sierra Leone			27●			41●				25△
										88●
Sudan					373●		638▲	174●	502●	245●
Togo					170●		92●		114●	55●
						68▲			74▲	
Turkey					2 640▲		3 100▲			
				1 223■	873■	2 600■				
Uganda							56●	22●		
Uruguay									815▲	
Venezuela										20 750△
Yugoslavia									988b●	500b●
									1 586▲	1 246▲
Zaire		211●	236●		1 147●	402▲	574●		1 317▲	
Zambia									285●	150●
										75△
TOTAL	373	1 350	373	1 806	6 179	3 723	5 757	2 382	51 089	116 220

Legend:
● Paris club renegotiation
▲ Commercial bank renegotiation
■ Aid consortia renegotiation
△ Agreed in principle

Figure 2.7 Multilateral debt negotiations, 1975–1984 (US$ million)
Data in italics are estimates.
a, An agreement of a special task force; b, agreement of a creditor group meeting, not a Paris Club; c, includes debt of US$23 625 previously rescheduled in 1983; d, proposed.
Source: World Bank (1985b, 28)

the price which they exacted for the IMF's direction that they must lend involuntarily to developing country debtors. The banks' second riposte to the IMF was to seek sufficient new business elsewhere that involuntary loans to problem debtors could be substantially reduced in the future. By 1985 this strategy was showing signs of success. New commercial bank lending to Latin America dropped from $19.182 billion in 1983 to $8.704 billion in 1985 (and $6.708 billion in 1986).

Net transfers

The decline in new bank lending which set in from 1983 had obvious consequences for the direction of net transfers to and from most debtor countries. A basic assumption of the containment strategy was that involuntary bank lending would give way to voluntary bank lending as a temporary series of country-specific liquidity crises was resolved. By 1985 this assumption was open to question and the possibility of a moratorium on all debt payments was being discussed in some Latin American countries.

The two issues are closely linked. Most students of the debt crisis recognize that there are compelling reasons which caution a debtor country against a repudiation of its debts. It is not just that such a default, if generalized, would threaten the international financial system, and thereby the financial assets of the well-to-do in the indebted world; it is also the case that the costs of default would comprise an uncertain but positive degree of creditor country retaliation. The US government could not be expected to sit idly by while its major commercial banks were thus abused or 'cheated'. The USA would act to seize the assets of defaulting countries in the USA (and elsewhere by agreement). It would also revoke short-term financing facilities and endeavour to exclude the defaulting nation or nations from the world economy. Military action would remain as an unlikely but possible last option.

Away from the arena of geopolitics, a strategy of repudiation has been discouraged on less threatening grounds. Possible repudiators have been told that such an action will prevent them from gaining access to official and private loans when the world economy recovers. The threat of no jam tomorrow is then used to discourage a strategy of seeking some jam today. Not everyone accepts the

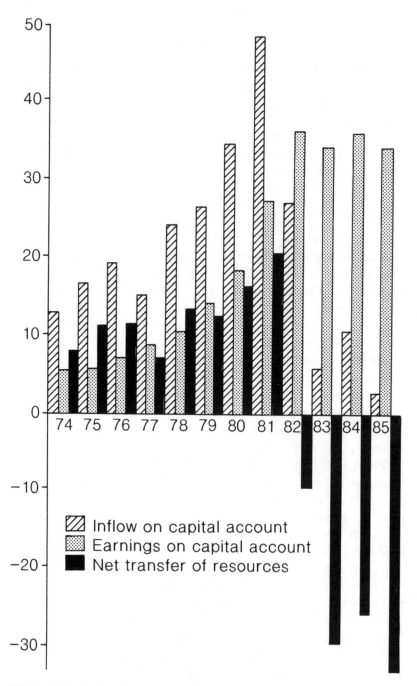

Figure 2.8 Latin America: capital transfer and net transfer of resources, 1974–1985 (US$ billion)
Source: Adapted from Roddick (1988, 15)

logic of this argument, however. For one thing there is little evidence to suggest that defaulting countries are punished at a future date (Lindert 1989). Banks have short memories, and new governments can and do take on different colours to old governments. Banks tend to look to the present and future, and not to the past. More pertinently, the jam tomorrow argument seems to ignore the pains caused by adjustment through austerity today. This argument has played especially well in Peru, where, in July 1985, President Alan Garcia 'announced his decision to limit Peru's foreign debt service payment to the equivalent of 10 percent of annual exports' (Roddick 1988, 169). Garcia tapped a consistent populist line when he called for a fairer sharing of the burdens of adjustment. Not without justification, he pointed out that the banks had not suffered a great deal as a result of the debt crisis in Latin America. By contrast, the peoples of Latin America had suffered visibly. This much was evident from a reading of the main development indicators of the region (see later sub-section). It could also be inferred from the size of the net transfers of funds from Latin America and the Caribbean between 1982 and 1985 (see figure 2.8; not based on World Bank data). The transfer of resources away from the developing world made the option of repudiation less abstract and remote. If actions could be taken to establish a debtors' cartel, and assuming that net transfers would not be positive for some time, repudiation might come to be seen as a reasonable and 'developmental' option. This fact alone might be sufficient to make the creditors think seriously about the wisdom of persisting for too long with a simple containment-austerity strategy of debt management.

Trade

As it turned out, there were other reasons which caused the creditors to think again about their debt management strategies. One such reason concerned an emerging debt overhang; another concerned the question of international trade.

Evidence on the trading element of the debt containment strategy is mixed. Most indebted regions did improve the position of their current accounts between 1982 and 1985 (see figure 2.9). In Latin America, Brazil and Mexico were especially praised for their efforts. In Brazil a current account balance of minus 16,312 million in 1982 was turned into a positive balance of $42 million in 1984, and only a slight negative balance (minus $273 million) in 1985. The corresponding figures for the trade account are $778 million, $13,086 million and $12,466 million. In Mexico a current account deficit of $13,899 million in 1981 was converted to a current account surplus of $5403 million in 1983 (and a smaller surplus of $1130 million in 1985). The corresponding figures for the trade balance are minus $4099 million, $13,762 million and $8451 million. Meantime, in Nigeria a current account deficit of $5854 million in 1981 was improved to a surplus of $2623 million in 1985.[6]

It is doubtful that such countries could have serviced the interest payments on their external debts without running a trade surplus in this manner. Nevertheless, such apparent success was bought at a price. Closer inspection of the trading statistics reveals that most of the adjustments made between 1982 and 1985 were made in terms of imports and not exports. For most indebted nations the option of exporting more was not an easy one. Exports of goods and services could be increased, but the process could also be thwarted by supply-side rigidities and by a creeping protectionism in the developed world. Thus, while a country like Brazil found itself in a not too difficult position (half of its exports being of manufactures), other countries (including Chile and Bolivia) were running to stand still as an increased volume of commodity exports fell foul of declining commodity prices. In such a situation, the easier option was to cut imports, which is what most debtor countries did between 1982 and 1985 (South Korea was a notable exception). In Mexico between 1982 and 1986 merchandise imports dropped by 20 per cent in value terms, while merchandise exports increased by 2.3 per cent from 1982 to 1985 (only to fall in 1986). In Brazil, the value of merchandise imports fell by 32 per cent between 1982 and 1985, while the value of merchandise exports rose by 27.1 per cent. In Nigeria the corresponding figures (1982 to 1985) are 42.9 and 5.9 per cent; in Indonesia, 28.8 and minus 6.2 per cent.

It matters too where the adjustments in a country's trading position are coming from. In Chile most of the transactions on the current account in this period involved public sector organizations. The foreign exchange then earned could be transferred without difficulty to debt service payments. In Brazil this transfer was less

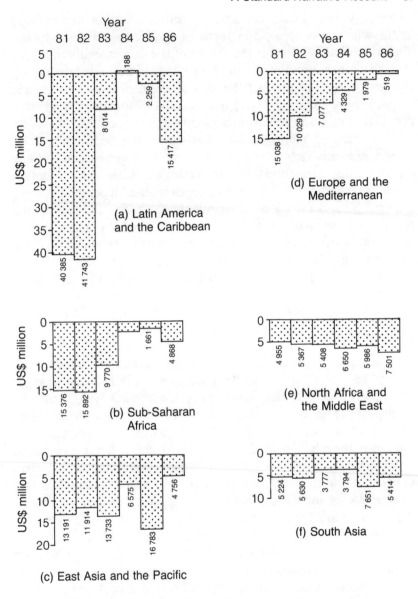

Figure 2.9 Current account balances of developing country regional groupings, 1981–1986
Source: World Bank (1989a)

assured, with the bulk of export earnings accruing to a private sector which was sometimes loath to transfer its largesse to the Treasury. Finally, there is the matter of the sustainability of such a trading strategy. On the one hand there is the question of the fallacy of composition, or the fact that not all countries can run a trade surplus at the same time (and not all debtor countries if a sizeable proportion of trade is between indebted countries). At the same time, a question mark can be raised about the desirability of an import-reducing debt servicing strategy. This option can become self-cancelling if resources are not available to make such an option viable over a longer period. Future exports often depend on current imports. This is the Catch 22 that lies at the heart of the acclaimed success of the trading adjustments witnessed in many debtor countries between 1982 and 1985. It is a paradox which is given some substance by the worsening debt/export ratios that characterized the period 1985–7 in most of Latin America and sub-Saharan Africa.

The debt overhang

A similar paradox is evident in the idea of a debt overhang. If the trading surplus strategy was designed to work in the short run, so too was it assumed that involuntary bank loans would only be needed for a short period of time. We have seen already how optimistic this assumption was. Most banks took steps to minimize the loans they had to make available to debtor countries on an involuntary basis, and most took steps to re-build their capital bases in other markets. They were further encouraged in this business strategy by an increasing fear of debt repudiation and by a belief that involuntary loans would be the last to be honoured if further defaults or debt moratoria became widespread. In other words, the previous build-up, or overhang, of debt argued against an infusion of new money. Paradoxically, this failure to advance new monies seemed likely to make the prospect of repudiation a reality. After all, an economy cannot grow if it is starved of investment funds and this is precisely what was happening to most debtor countries in the containment years. Monies which might otherwise have been invested in local industries (including export-oriented industries) were available only or mainly to support an immediate net transfer of funds away from the debtor country or region. (This transfer was much expanded by capital flight, on which more later.) Put bluntly, the short-term success of the containment-austerity strategy seemed to sow within it the seeds of its own demise. Rates of investment in Latin America dropped sharply between 1980 and 1984, as can be seen from table 2.4. For the region as a whole the rate of gross domestic investment fell by 33.8 per cent, with only Colombia standing out against this depressing trend. Meantime, in sub-Saharan Africa, the rate of gross domestic investment fell by 22.9 per cent between 1982 and 1985; in the Philippines it fell by 53.8

Table 2.4 Latin America: the fall in investment

Gross domestic investment, US$ billions[a]

	1980	1984	1984/1980 change(%)	1986
Brazil	66.4	41.9	−36.9	46.3
Mexico	51.5	34.2	−33.6	35.3
Argentina	18.3	9.0	−53.0	8.7
Venezuela	14.8	10.0	−32.4	12.1
Chile	6.5	4.0	−38.5	4.2
Peru	4.0	3.3	−17.5	3.1
Colombia	6.5	7.1	9.2	6.2
Ecuador	3.0	2.0	−33.3	2.2
Uruguay	1.6	0.8	−50.0	0.6
Bolivia	1.0	0.7	−30.0	0.6
Latin America	187.1	123.8	−33.8	125.7

Total investment as % of GDP

	1980–1 (average)	1982	1983	1984
Argentina	21.0	19.0	16.2	14.7
Brazil	23.5	24.3	20.8	17.9
Cile	22.7	12.3	10.4	14.9
Colombia	20.0	20.9	19.8	19.4
Mexico	29.4	22.5	18.9	18.5
Peru	21.0	23.9	20.8	17.1
Venezuela	23.7	26.5	14.7	16.0
Average of 7 countries	23.0	21.3	17.4	17.0

[a]In constant 1986 dollars.
Source: After Branford and Kucinski (1988, 8)

per cent, and in Indonesia it fell by 5.4 per cent. The rate of gross domestic investment in South Korea increased by 9.5 per cent between 1982 and 1985 (IMF *Yearbook of International Financial Statistics* 1988).

Development

The effects of the containment strategy on development are not hard to guess at. In the early 1980s most African and Latin American countries began to 'underdevelop'. This much is obvious from a reading of the most basic macroeconomic indicators: GDP, GDP per capita, the position of international reserves, etc. (see table 2.5). It could also be deduced from the massive rise in unemployment which hit the formal sector of most indebted nations (see table 2.6 for data on Latin America). This rise in unemployment meant that households which previously had been able to cope with poverty were now left exposed and vulnerable. The whole unhappy edifice was then capped in many countries by rampant inflation, as governments sought to minimize the political impact of austerity measures by borrowing at high interest rates and/or by printing money. In Brazil the annual inflation rate in 1985 was 242.2 per cent; in Bolivia, in 1985, the rate was estimated to be over 8000 per cent. In such circumstances, precise figures mattered very little.

Not all countries and regions, or classes and households, suffered in the same way, and attention will be given to these and other distinctions in due course. Nevertheless, it is difficult to end this brief review of the debt containment years without echoing the words of Pedro-Pablo Kuczynski, ex-Minister of Energy and Mines in Peru and later chairman of First Boston International. In his judgement: 'By 1985 only the austerity component of the containment strategy seemed to be firmly in place; new financing and export growth appeared on their way to failure' (Kuczynski 1988, 94). In Latin America, development had been sacrificed to secure the stability of the international banking system.

Table 2.5   Selected economic statistics for Latin America, Asia and Africa, 1981–1985

GDP (constant prices): percentage change over previous year					
	1981	1982	1983	1984	1985
Africa	−0.4	−1.3	−2.5	−0.1	2.6
Asia	5.9	5.2	6.6	4.6	4.5
Latin America and the Caribbean	−0.1	−1.5	−2.3	−3.5	2.5
Brazil	−3.3	0.9	−2.5	5.7	8.3
Mexico	7.9	−0.6	−5.3	3.7	2.8
Argentina	−6.7	−5.0	2.9	2.5	−4.4
Venezuela	−0.3	0.7	−5.6	−1.4	0.3
Nigeria	−8.4	−3.2	−6.3	−5.2	5.3
Côte d'Ivoire	2.0	3.0	−12.9	−	−
South Korea	7.4	5.7	10.9	8.6	5.4

GNP per capita (constant prices; US dollars)			
	1981	1983	1985
Brazil	2220	1880	1640
Mexico	2250	2240	2080
Argentina	2560	2070	2130
Venezuela	4220	3840	3080
Nigeria	870	770	800
Côte d'Ivoire	1200	710	660
South Korea	1700	2010	2150

Total reserves minus gold (millions of SDRs: end-period)					
	1981	1982	1983	1984	1985
Brazil	5673	3561	4160	11 740	9654
Mexico	3500	756	3737	7419	4467
Argentina	2808	2272	1120	1268	2844
Venezuela	7014	5964	7300	9081	9332
Nigeria	3347	1462	946	1492	1518
Côte d'Ivoire	15	2	19	5	4
South Korea	2304	2545	2241	2809	2612
Africa	10 310	7100	6785	6612	8219
Asia	30 493	35 470	43 346	51 483	45 804
Latin America and the Caribbean	32 821	24 476	26 865	40 529	36 652

Sources: IMF, *International Financial Statistics Yearbook*, 1987; World Bank, *World Development Report*, 1983, 1985, 1987; ECLAC, *Economic Survey of Latin America and the Caribbean*, 1986, 1988

Table 2.6 Latin America: open and hidden unemployment in the urban economy, 1970–1985

	1970		1980			1983			1985		
	Urban informal sector (%)	Open unemploy ment (%)	Urban informal sector (%)	Open unemploy ment (%)	Combined change[a] 1980/1970 (%)	Urban informal sector (%)	Open unemploy ment (%)	Combined change[a] 1983/1980 (%)	Urban informal sector (%)	Open unemploy ment (%)	Combined change[a] 1985/1980 (%)
Latin America[b]	29.1	6.8	26.1	6.9	-2.9	29.0	10.2	+6.2	30.7	11.1	+8.8
Chile	22.9	4.1	36.1	11.7	+20.8	37.2	19.0	+8.4	37.2	17.0	-0.9
Costa Rica	22.5	3.5	28.6	6.0	-8.6	29.3	9.9	+4.6	28.3	6.7	-0.4
Argentina	19.0	4.9	26.3	2.6	-5.0	27.1	4.6	+2.8	28.9	6.5	+6.5
Peru	39.8	8.3	34.2	10.9	+3.0	32.7	13.9	+2.4	34.9	17.6	+7.4
Colombia	30.9	10.8	30.2	9.7	-1.8	33.6	11.8	+5.5	35.4	14.1	+9.6
Brazil	27.5	6.5	24.1	6.2	-3.7	29.6	6.7	+6.0	30.1	5.3	+5.1
Venezuela	30.7	7.8	25.6	6.6	-6.3	27.3	10.5	+5.6	26.2	14.3	+8.3
Guatemala	43.4	–	31.5	2.2	-11.9	32.8	7.6	+6.7	33.5	12.9	+12.7
Mexico	34.2	7.0	24.2	4.5	-13.0	25.6	6.7	+3.6	29.5	4.8	+5.6

Unemployment is given as a percentage of the non-agricultural labour force.
[a]Increase in open and hidden unemployment together.
[b]Arithmetical average.
Source: Roddick (1988, 92)

7   The Baker Plan: Adjustment with Growth

The mixed success of the containment strategy was apparent not only to academics and to citizens of the indebted nations. On 8 October 1985, the Secretary of the US Treasury, James Baker III, announced a new debt initiative. The venue for this change of direction was Seoul in South Korea, the occasion being the annual Joint Meeting of the IMF and the World Bank. After first thanking his hosts and praising South Korea, whose 'market-oriented approach and strong emphasis on private initiative are a lesson for us all', Secretary Baker acknowledged that the industrial countries would have to take up some of the burden of adjustment in the developing countries' debt crisis. More exactly, Baker acknowledged the successes of the containment strategy, and he noted that the strategy had become more flexible with the introduction of MYRAs in 1984. He particularly endorsed the suggestion that MYRAs should be offered 'as rewards for countries that had made strong progress on policies to deal with their balance of payments problems' (World Bank 1989a, xviii). At the same time, he proposed that: 'If the debt problem is to be solved, there must be a Program for Sustained Growth incorporating three essential and mutually reinforcing elements –

1   First and foremost, the adoption by principal debtor countries of comprehensive macroeconomic and structural policies, supported by the international financial institutions, to promote growth and balance of payments adjustment, and to reduce inflation.
2   Second, a continued central role for the IMF, in conjunction with increased and more effective structural adjustment lending from the multilateral development banks (MDBs), both in support of the adoption by principal debtors of market-oriented policies for growth.
3   Third, increased lending by the private banks in support of comprehensive economic adjustment programs' (Baker 1985).

Thus was born the 'Baker initiative' and with it a new optimism on the debt crisis. The watchwords now were *adjustment with growth* and the promise seemed to be that new resources would be made available to some debtor nations (see also Selowsky and van der Tak 1986). In part, these resources would be delivered by a 'serious effort to develop the programs of the World Bank and the Inter-American Development Bank . . .[to] increase their disbursements to principal debtors by roughly fifty per cent from the current annual [1985] level of nearly $6 billion' (Baker 1985). Within this expanded program particular increases in resources were to be earmarked for the International Finance Corporation of the World Bank Group, and for the Bank's Multilateral Investment Guarantee Agency. Both organizations, declared Secretary Baker, 'can do much to assist their members in attracting non-debt capital flows as well as critical technological and managerial resources' (ibid.). Other resources would flow in the form of FDI and as renewed and expanded voluntary private bank lending. At the end of his address, Secretary Baker suggested that the commitment 'by the banks to the entire group of heavily indebted, middle income developing countries would be net new lending in the range of $20 billion for the next three years'. He concluded:

> I would like to see the banking community make a pledge to provide these amounts of new lending and make it publicly, provided the debtor countries also make similar growth-oriented policy commitments as their part of the cooperative effort. Such financing could be used to meet both short-term financing and longer-term investment needs in the developing countries, and would be available, provided debtors took action and multilateral institutions also did their part. (Baker 1985)

The Baker initiative was warmly received by the Western press. In February 1985 *Fortune Magazine* had run an article by Gary Hector entitled: 'Third World Debt – The Bomb is Defused'. Hector claimed that 'Evidence is building that the international debt crisis is over' (Hector 1985, 24). Although 'A feeling of hard times. . .still pervades the major Latin American countries', on balance 'confidence is growing in the debt-restructuring process', and in the booming Latin American exports which were a complement to 'the US economic boom' (ibid.). Now the Baker intitiative seemed to confirm such optimism (Morgan Guaranty Trust 1986). The context for Secretary Baker's address was a US economy at full capacity; a confident, Reaganite economy which seemed to have delivered a miracle to US citizens and which could now act as a locomotive for the rest of the world (and for indebted Latin America in particular). The time was right to promise a less austere future for some indebted nations; to hold out the prospect of less protected markets in the industrialized world, and to dangle the carrot of new bank loans as a reward for a continuing process of structural adjustment.

Before an asessment of the Baker years is offered, it is important to note that the rhetoric which surrounded the Baker intiative can disguise a greater and underlying continuity in the creditor countries' debt management strategy. It is significant that the 1985 proposals did not raise the possibility of non-market write-downs in the debt and debt service owed by Latin American countries. Moreover, the Baker initiative continued to be a Latin American initiative; a banking initiative as much as a debt initiative. When Baker referred to the heavily indebted middle-income countries, he had in mind 15 countries in particular (the highly indebted countries, minus Jamaica and Costa Rica). In 1986–7 this group of 15 was expanded to 17, but its regional composition remained more or less unaltered: 12 countries from Latin America, one from Europe, one from East Asia, one from North Africa and two from sub-Saharan Africa. Low-income countries heavily indebted to official creditors were not mentioned in Baker's address. Finally, Baker was keen

> to emphasize that the United States does not support a departure from the case-by-case debt strategy we adopted three years ago. This approach has served us well; we should continue to follow it. It recognizes the inescapable fact that the particular circumstances of each country are different. Its main components, fundamental adjustment measures within the debtor nations and conditionality in conjunction with lending, remain essential to the restoration of external balance and longer-term growth.

Assessment

The fact that the Baker initiative was largely replaced by the Brady initiative in 1988–9 suggests that it was not a great success; either that, or it was so close in design and execution to elements of the first and third adjustment strategies that it is difficult to evaluate as a distinctive contribution. Nevertheless, certain points can be made with confidence about the years 1985–7. These points concern the following subjects: world economic growth and regional economic linkages; continuing structural adjustment and net new bank lending; and the performance of the major development indicators.

Economic growth

The years 1985–7 were years of relative stability and rapid growth in the world economy, at least in the industrial countries. The US dollar reached a peak in September 1985 and thereafter was driven down in accordance with the Louvre accords of that time; meanwhile inflation averaged only 4 per cent in the economies of the OECD, while the rate of unemployment in the G7 countries had come down from an average of 8.25 per cent in 1983 to an average of 7.25 per cent in 1987. Interest rates were lower in both nominal and real terms than they had been in the first half of the 1980s. The World Bank, pleased with this state of affairs, introduced its World Development Report for 1985 with the claim that 'The economic turbulence of the past few years has subsided' (World Bank 1985b, 1).

It would be reasonable to assume that the buoyancy of the 'world economy' between 1985 and 1988 would have beneficial consequences for the indebted nations. One clear expectation of the Baker initiative was that, 'as growth resumed and export earnings increased, debt ratios would progressively fall. This, in turn, would gradually restore creditworthiness and pave the way for voluntary commercial lending' (World Bank 1989a, xix). In practice, no such chain of events became apparent. It may be that the indebted nations would have fared less well in a less buoyant 'world economy' between 1985 and 1988. The positive linkages, however, are more difficult to detect. As table 2.7 makes clear, the performance of the Baker 17 countries was, at best, mixed. In terms of GDP growth some countries, including Chile, Brazil and Colombia, performed relatively well (with average annual rates of growth of 4 per cent between 1982 and 1988, not allowing for population growth). Other countries performed dismally, as, for example, did Jamaica, Peru, Nigeria, Yugoslavia, Uruguay and even Mexico. In terms of exports and imports the story was similarly mixed, with many countries being hit in 1987 by a sharp fall in non-oil commodity export prices. (The World Bank makes reference to a terms of trade loss of 21 per cent in the two years 1986 and 1987.) Mexico and Nigeria were also damaged by a sharp fall in world oil prices in 1986. More generally, while the HICs recorded 'combined annual trade surpluses of the order of $30 billion. . .their aggregate current account, which was positive in 1984 and 1985, again moved into deficit in 1986 and 1987' (World Bank 1989a, xix). The export-to-growth strategy was not paying off. Finally, in Latin America and the Caribbean as a whole, the ratio of gross investment to GDP fell from '21.4 per cent in 1980 to 14.9 per cent in 1987' (World Bank 1989a, ix). In some countries, 'current investment ratios [were]. . .barely sufficient to cover depreciation' (ibid.). Notwithstanding some possible improvements in the efficiency of use of existing capital, the World Bank felt obliged to acknowledge that 'A persistence of the low investment ratios reached in 1987 will also continue to hold back future growth' (ibid.).

It is not easy to say why so many indebted countries fared so poorly between 1985 and 1987. Continuing capital flight and domestic economic mismanagement may be partly to blame, as many creditors were quick to suppose. Just as important, however, were issues relating to supply-side bottlenecks, industrial country protectionism and the nature of international economic multipliers. In each case the Baker initiative had tended to assume that growth in one part of the world would lead inexorably to rapid growth in other regions. This did not happen, and nor was it likely to given the continuing drain of net transfers from the indebted countries, given the damaging effects of a mounting debt overhang, and given the non-textbook functioning of international trading systems. The positive multipliers which could be observed between the three parts of the OECD 'world economy' – the USA, the EC and Japan – were less evident with regard to North–South linkages and other assumed interdependencies. In any case, such positive multipliers could easily be offset by further interest rate rises (the World Bank estimates that close to 1 per cent of Latin America's GDP during the 1980s was consumed by high real interest rates; World Bank 1990b, 15), and/or by further increases in the non-tariff barriers which had helped to depress developing country export earnings.

Table 2.7 Debt, debt service and growth in the 'Baker 17' countries

| Country | Debt outstanding 1988[a] | | Debt service 1988–90[b] | | Debt ratios | | Average annual growth rates[d] 1982–81 (%) | | | Invest-ment (%) | Per capita con-sumption (%) |
	Total	Of which private source (%)	Total	Of which interest	DOD/ GNP 1987	Interest/ XGS 1987[c] (%)	GDP	Exports	Imports		
Argentina	59.6	79.4	17.7	11.4	73.9	41.5	1.4	1.9	1.3	-2.1	-0.4
Bolivia	5.7	27.3	1.8	0.8	133.7	44.4	-1.4	-1.3	5.6	-16.7	-1.6
Brazil	120.1	76.8	63.4	21.8	39.4	28.3	4.8	4.2	-2.0	2.8	2.6
Chile	20.8	74.3	7.0	5.2	124.1	27.0	4.3	7.1	3.6	15.1	-0.8
Colombia	17.2	48.0	10.3	3.6	50.2	17.0	4.1	9.4	-1.7	-0.1	1.3
Costa Rica	4.8	53.2	2.2	0.7	115.7	17.5	3.6	1.4	8.3	9.3	2.6
Côte d'Ivoire	14.2	60.2	5.0	2.2	143.6	19.7	1.3	-1.0	-4.8	-9.0	-2.1
Ecuador	11.0	63.6	5.5	2.1	107.4	32.7	1.5	5.6	-2.3	-2.1	-2.4
Jamaica	4.5	17.6	1.6	0.7	175.9	14.2	0.7	10.8	8.8	-2.2	-0.3
Mexico	107.4	78.1	43.5	24.0	77.5	28.1	0.2	4.1	-1.0	-4.5	-1.8
Morocco	22.0	29.0	9.7	2.9	132.4	17.3	3.6	5.4	0.5	-0.7	0.9
Nigeria	30.5	61.1	16.4	4.6	122.6	23.3	-0.3	2.1	-20.6	-10.1	-4.5
Peru	19.0	61.5	7.4	2.4	40.5	27.2	2.9	-1.1	-6.0	-11.9	-1.4
Philippines	30.2	60.0	11.9	5.0	86.5	18.7	-0.1	5.5	2.4	-12.0	-0.6
Uruguay	4.5	77.1	1.8	0.8	58.6	17.7	1.7	4.7	1.2	-3.4	1.0
Venezuela	35.0	99.3	15.6	7.8	94.5	21.9	1.2	1.2	-1.4	-1.6	-1.4
Yugoslavia	22.1	61.9	13.8	4.4	38.9	10.8	1.0	1.0	-1.7	0.2	0.3
Total	528.6	71.2	234.6	100.2	63.1	24.2	2.6	2.9	-3.4	-1.5	0.2

Values are in US$ billions unless otherwise indicated

[a] Estimated total external liabilities, including the use of IMF credit.

[b] Debt service is based on long-term debt at end 1987. It does not take into account new loans contracted or debt reschedulings signed after that date.

[c] Based on interest due in 1988 on long-term debt outstanding at the end of 1987, relative to 1987 exports of goods and all services.

[d] Data for 1988 are preliminary estimates. Growth rates (least squares) are computed from time series in constant prices.

Source: World Bank (1989a)

Structural adjustment and new net bank lending

A second weakness in the Baker initiative concerned the different powers which the multilateral organizations could bring to bear upon indebted client nations and private commercial banks. The IMF and the World Bank were able to secure a continuing structural adjustment in most debtor countries, notwithstanding an often determined local opposition. The idea that the debt crisis was a punishment for fiscal laxity in the 1970s was pushed more forcefully in the Baker and Brady years; as, for example, when the World Bank declared that 'Protracted fiscal laxity and economic mismanagement in the 1970s had been among the main causes of the debt crisis' (World Bank 1989a, xix). The cure for such laxity was austerity and a rolling back of the financial claims made on the public purse.

When it came to new bank lending, however, the World Bank's power to coerce was less obvious, and less and less real. Most of the commercial banks had improved their operating positions by 1986–1987, and most looked with scepticism upon the Baker proposals for $20 billion of net new lending to proven bad debtors. Other areas of private bank business were more profitable, or at least more assured, and a growing debt overhang continued to deter new net lending on the grounds that bad old debts would be serviced before bad new debts. Less prospectively, the story of the banks' failure to commit significant new resources to the indebted countries is told by the World Bank. Net

> long-term lending by private creditors to the HICs. . .fell from $7.2 billion in 1984 to $1.3 billion in 1985 and to $−1.9 billion in 1986. Lower lending – which stood in contrast to the Baker initiative's recommendation that lending be increased – in turn caused further negative resource transfers from the HICs to their external creditors. During the [period] 1985–87 they amounted to nearly $74 billion – 3.1per cent of the HICs' combined GNP for that period and almost equal to the drop in investment. (World Bank 1989a, xix)

The story needs little in the way of elaboration.

Development

The prospects for development remained bleak in the Baker years. Although Colombia and South Korea proved themselves exceptions to the rule, a combination of net resource transfers, capital flight, low investment and austerity for the most part secured predictable results. In the years 1985–7 inclusive, GDP per capita increased by just 3.4 per cent in Latin America and the Caribbean (ECLAC 1989), and fell by 18 per cent in sub-Saharan Africa (IMF 1988). Unemployment rates (which are variously constructed and which are sometimes no more reliable than inflation statistics) increased throughout the indebted developing world, albeit at different rates in different countries.

Within this broad picture, particular regions and classes fared more or less badly. Although little work has been done on this topic, it would be too simple to argue that it was only the poor and most vulnerable groups who were hurt by the effects of structural adjustment. In Mexico, there is evidence to suggest that middle-class, public sector workers suffered the sharpest relative decline in living standards (Lustig 1990). It remains the case, however, that such groups had some distance to fall; households and communities already on the margin could find themselves in worsening poverty on account of a very small change in their fortunes.

Susan George documents this very well in her book, *A Fate Worse than Debt*. It is worth closing this section with a passage from that volume. In this passage, George recounts the plight of Bolivian mother, Zona San Jose Carpinteros, amidst the debt-induced evils of rising inflation and unemployment. Says the mother:

> Since everything is so expensive, I don't give my children breakfast any more. For lunch I give them a little rice soup. I don't buy sugar now that it has gone up. To eat, I have to make do any way I can, because the children can't get along without food. Us adults, we manage without when we have to. Sometimes I say to myself, 'I'm going to give away my children to someone'. But then I think of what my parents might do to me – that's what I'm afraid of. (George 1989, 147)

This is what a 'lost decade of development' often means at the level of an individual household.

8 Brady and the Market-menu Approach

The Baker years were a prelude to more radical proposals for debt management, in both the HICs and sub-Saharan Africa. In 1988 and 1989 the Toronto and Brady initiatives were announced and the containment strategy was quietly retired in favour of an approach which required the market write-down of debt and debt service, and a continued rescheduling of debt repayments.

The mainsprings of this 'second transition' are to be found within the Baker initiative, in terms of both its mandate and its limitations. By 1987 it was apparent that new net bank lending to the HICs was not forthcoming in the manner called for by James Baker. It was also apparent that most banks were taking steps to reduce their debt exposures by increasing their loan-loss reserves. (In part, this was in anticipation of the Basle Agreement on Capital Adequacy Guidelines; in part it was in line with a speech by Mr Baker to the 1987 joint Annual Meetings of the IMF and World Bank, which listed an expanded range of financing options for commercial banks involved in rescheduling agreements.) Some banks were also seeking to exchange existing debt for new assets in the emerging swaps and secondary markets. In 1988, 'the stock of commercial banks' claims on the Baker 17 countries fell by $19.2 billion. Claims of banks reporting to the Bank for International Settlements on Mexico declined by $6.5 billion during 1988 alone, and commercial bank claims on Brazil fell in 1988 by $5.1 billion' (World Bank 1990a, 15). Debt was increasingly being concentrated in the non-private sector.

In 1989 this new attitude on the part of the private banks was accounted a virtue, instead of being condemned as a vice. Political circumstances were now such that an end to austerity in Latin America was supported by a powerful constituency of exporters in the USA. The World Bank reported that: 'The US House and Senate proposed bills in 1987 that included specific statements on the adverse effects of the debt problem on US exports and of the increase in imports from debtor countries by the United States' (World Bank 1990a, 17). The academic and policy-related rhetoric also changed. The banks were now praised for their willingness to engage in market-based debt write-downs. With not a little irony, it was suggested that they were acting in the spirit of the Baker intitiative. The sub-text, after all, of the Baker speech of 1985 was that '[domestic] policies should consist of market-oriented exchange rate, interest rate, wage, and pricing policies to promote greater economic efficiency and responsiveness to growth and employment opportunities; and sound monetary and fiscal policies focused on reducing domestic imbalances and inflation and on freeing up resources for the private sector.' The banks were articulating this philosophy in a specific arena. They were also acting on a case-by-case basis and clearly they were not opposed to a continuing structural adjustment in the developing countries. Instead of being cast as the villains of the piece, the banks were now hailed as the progenitors of a new dawn for creditor–debtor relations.

This new dawn found a formal voice in a speech made by Secretary to the US Treasury Nicholas Brady to the Bretton Woods Committee on Third World Debt in March 1989. (It also coincided with reports that 'In Argentina and Venezuela, social unrest was evoked by austerity measures associated with the countries' external debt burden' (World Bank 1990a, 17) and that arrears to creditors had risen to $52 billion in December 1988. LIBOR was also beginning to rise again following the stock market crash of October 1987.) Following earlier speeches by President Mitterand of France, Chancellor Lawson of the UK, Japanese Finance Minister Miyazawa, and the Chairman of American Express, James Robinson, Secretary Brady was minded to build upon the officially sanctioned Bolivian debt buy-back scheme of 1988 (of which more later) and other well publicized developments in the secondary debt markets. Having made a ritual declaration that 'the fundamental principles of the current [Baker] strategy remain valid', Mr Brady suggested that 'the path toward creditworthiness of severely indebted countries should involve debt and debt service reduction on a voluntary and case-by-case basis, in addition to rescheduling of principal and new money packages' (World Bank 1990a, 21). The main difference between the Brady initiative and 'the *ad hoc* debt reduction that had taken place until then through[a] market-based menu strategy was its inclusion of official support for debt and debt service reduction. The World Bank and the IMF were asked to provide funds for debt and debt service reduction operations for countries with high external debt burdens and strong adjustment programmes' (World Bank 1990a, 21).

The Brady proposals were directed mainly at the SIMICs, as the severely indebted middle income countries were now labelled. The package was adopted by the IMF and the World Bank in May 1989 when both institutions 'adopted operational guidelines and procedures for debt and debt service reduction' (World Bank 1990a, 21). These guidelines and procedures made it possible for SIMICs which had carried out medium-term adjustment programmes acceptable to the World Bank and/or the IMF to apply for resources from these two organizations to support local market-based proposals for debt and debt service reductions. The expectation was that the IMF and the World Bank would commit resources to the programme of the order of $20–25 billion over a three-year period from 1989. Half of this lending would come from existing lending programmes, with the rest being in the form of additional lending to provide interest support on exchanged debt. Finally, it was hoped that Japan would boost the Brady proposals by committing up to $10 billion in cofinancing over a period of three to four years. This was duly secured.

The Brady initiative has to date comprised several country programmes for debt and debt service reductions, and several associated innovations in respect to the market menu of techniques available for officially supported debt and debt service reductions.

Private Debts

Taking the country programmes and techniques in tandem, agreements were first reached with Mexico, the Philippines and Costa Rica, with later deals being agreed with Venezuela, Morocco and Uruguay. The first agreements were to be concluded by October 1990. While all these programmes are on a case-by-case basis, they share a commitment to use a market-based menu of options which are entered into on the basis of prior agreements between the debtor countries and their Bank Advisory Committees.

In Mexico, the focus of the agreement was on debt reduction instruments designed to cut the net cost of Mexico's commercial debt. The agreement asked 'banks to take a loss on present exposure of about 35 per cent or recycle about two-thirds of interest

payments over [a] four year [period]. This structure [was] designed
to provide a predictable flow of financial relief to Mexico over the
next several years and to reduce the uncertainty surrounding its
external public finances' (World Bank 1990a, 23). The commercial
banks were also encouraged to lend new monies to Mexico, and to
choose different combinations of debt and debt service reductions
from a local menu of options. The expectation was that Mexico's
net transfer abroad between 1989 and 1995 would be reduced by
about $4 billion per annum as a result of the Agreement, with about
$8 billion (or 8 per cent) of Mexico's net external debt being
reduced in addition. The indirect effects of the Agreement with
Mexico were expected to be felt in terms of lower domestic interest
rates and a greater willingness of foreign corporations to invest in a
stable nation. The participation of the private banks in the Agreement was rightly assumed to be conditional on the nature and scale
of official support for the proposed processes of debt and debt
service reduction.

The Philippines Agreement was quite different to that signed
with Mexico, being in essence 'a new money agreement with an
extended buyback operation' (World Bank 1990a, 23). A buyback
involves an indebted country buying back its debts at a cash
discount. Under the terms of the Brady initiative such debt
buybacks were to be officially supported, with the IMF and the
World Bank helping to provide the necessary finances for the
operation. The rate of discount on the debt would be established by
reference to the secondary markets which had (re-) emerged in the
early 1980s and which had expanded rapidly after 1986 (see table
2.8). The secondary market did not begin with the sort of debt
buyback pioneered by Bolivia in 1988. The market grew rather by
virtue of swaps between commercial banks, where two or more
banks were seeking a different portfolio of debt risks. The market
became more significant from 1985 when Chile and Mexico introduced 'debt conversion programs to reduce outstanding external
claims. . .based on swaps that involved the sale of claims at a
discount' (World Bank 1990a, 18). Most of these swaps took the
form of debt for equity swaps 'in which an investor purchases debt
in the secondary market at a discount and exchanges it for an equity
investment in the debtor country' (ibid.; see also Anayiotos and de
Piniés 1990; Ffrench-Davis 1990). Later conversions included

buybacks, informal private or quasi-private agreements (such as debt-for-nature swaps); debt for debt swaps (in which 'foreign currency debt is exchanged for obligations denominated in domestic currency'; World Bank 1990a) and exit bonds (where creditor banks seek to avoid future concerted lendings).

Finally, there is the Costa Rica Agreement. This settlement called for no new money from commercial bank creditors, but proposed instead a debt buyback at a substantial discount and a regularization of Costa Rica's arrears to commercial banks. As with the Mexico and Philippines agreements, the Costa Rica Agreement presupposed a continuing local commitment to structural adjustment and the deregulation of the domestic economy. The Agreements were also intended to run in combination with changes in the tax and regulatory regimes of the creditor countries. These changes offered tax incentives to banks which complied with the Brady initiative. This involved a certain amount of socialization of the debt in the USA and elsewhere, notwithstanding an avowed emphasis upon a market-menu of debt reduction agreements.

Table 2.8 Conversions of developing country external debt, 1984–1988 (US$ million)

	1984	1985	1986	1987	1988
Debt–equity swaps	773	1843	1522	3335	9205
Exit bonds	0	0	0	15	4275
Buybacks	0	0	0	0	648
Informal	0	0	0	3500	5414
Other	0	245	714	1337	2366
Total conversions	773	2088	2236	8188	22358
Memorandum item: total volume of debt swaps, including interbank trading	2000	4000	7000	12000	50000

Source: World Bank (1990a)

Official Debts

Mention of the role of creditor governments brings us to a second strand in recent debt management intiatives: to the role of the Paris

Club and the Lawson and Mitterand initiatives. More exactly, it brings us to most of the non-Latin American debtor countries and to a brief survey of how they fared and were treated in the 1980s.

Some background is needed first. Although mention is commonly made of an African debt crisis, or of a debt crisis in sub-Saharan Africa, the African countries vary in terms of their patterns of economic performance, debt stocks and flows, and debt management plans. Moreover, the African countries do not alone comprise the so-called SILICs, or severely indebted low income countries: Myanmar and Guyana also fall into this category (as of December 1990). These caveats entered, there is still some merit in referring to a sub-Saharan African debt crisis. To begin with, total debt stocks tend to be low in sub-Saharan Africa, but high in relation to local debt servicing capacities. At the end of 1989 the total external debt of sub-Saharan Africa was estimated at $147 billion, compared to a total in Latin America and the Caribbean of $422 billion. Relative to GNP, debt in sub-Saharan Africa in 1989 was 98.3 per cent, compared to 46.7 per cent in Latin America and the Caribbean. For IDA-only sub-Saharan Africa, the ratio of total external debt to exports of goods and services was consistently above 300 per cent in the 1980s. Second, the structure of debt outstanding in Africa is biased heavily towards official creditors. Nigeria is an exception to this rule (61.6 per cent of its outstanding long-term public and publicly guaranteed debt was owed to private creditors in 1988), but elsewhere about 90 per cent of debts outstanding in 1989 were owed to bilateral and multilateral official creditors. Third, the payments record of sub-Saharan Africa is less consistent than that of Latin America. This largely reflects the poverty of most of sub-Saharan Africa and a debilitating lack of infrastructure. Export prospects, too, have often been grim. Most economies are not diversified in terms of their export bases, and most have suffered from a decline in the terms of trade for their commodity exports since 1982. In this context arrears on debt are not surprising. Finally, many African debtors have been in repayments difficulties for a period of time substantially longer than ten years. Zaire rescheduled its debts on several occasions in the 1970s and 1980s, as did Liberia, Madagascar, Senegal, Sudan and Togo. In response to this, the African debt crises have been dealt with on the basis of continuing infusions of concessional finance and by means of debt reschedulings through the Paris Club of official creditors. During the 1980s the terms upon which debt was rescheduled tended to become more liberal, with longer maturities and grace periods being complemented by a progressive reduction in interest rates.

Since 1989–9, the debt crises of sub-Saharan Africa have been dealt with by means of a new menu of options (as per Brady and the SIMICs). The impetus for the new policy approach came from a series of speeches and interventions by individuals as diverse as Lawson, Mitterand and Willy Brandt. In practical terms the new approach comprises two main elements: a World Bank sponsored Special Program of Assistance for 23 low income (IDA) African countries, and a series of creditor responses to the African crisis which can be broken down by type of creditor (bilateral, multilateral and private). Details of each proposal are as follows.

(1) The Special Program for Africa. It is commonly assumed that Africa's debt problems will only be diminished on the basis of a medium- and long-term commitment to improved economic management within the region. Infrastructure has to be improved, incentives have to be provided in agriculture, and internal and external prices have to be more closely informed by the market. To aid this process, 22 debt-distressed low-income African countries are now eligible for support from the World Bank's Special Program for Africa (SPA), a programme initiated at a donors' conference in December 1987. The SPA 'endeavours to provide substantially increased, quick-disbursing, highly concessional assistance to adjusting countries' (World Bank 1990a, 31). It has five elements:

(1) increased adjustment lending from the eighth replenishment of the International Development Association (IDA 8); (2) increased cofinancing and coordinated financing from bilateral donors and other multilateral agencies, and (3) supplemental IDA adjustment credits to IDA-only countries with outstanding debt of the International Bank for Reconstruction and Development (IBRD). These resources are provided in conjunction with (4) additional concessional IMF resources from the Enhanced Structural Adjustment Facility, and (5) greater debt relief. Elements 3 and 5 are available to SILICs outside Africa. (World Bank 1990a, 31)

(2) Bilateral creditors. Following the Toronto summit of mid-1988, the Paris Club 'eased the terms on which official bilateral debt is rescheduled for severely indebted low-income countries over time' (World Bank 1991a, 93). In 1989–90, Paris Club reschedulings were available on a menu of options where option A involved the cancellation of one-third of the amount consolidated, option B involved very long maturities (25 years) and option C involved a reduction in interest rates. Creditor countries can choose their preferred option. In practice most countries have chosen options B and C, although France, Finland and Sweden have chosen option A.

Bilateral creditors can further relieve the debt crises of sub-Saharan Africa by increasing the flow of concessional aid to indebted countries. The World Bank notes with approval that 'sub-Saharan Africa has emerged during the last two decades as the major aid-receiving region. From less than 10 per cent of total ODA in 1960, it accounted for more than 30 per cent by the late 1980s' (World Bank 1991a, 93). A greater share of grant aid can be expected as part of the SPA arrangements (and as a result of the Trinidad Terms proposed in 1990 by UK Chancellor John Major).

Finally, bilateral creditors are encouraged to forgive a portion of the debt owed by some African countries. The forgiveness will take the form of cancelled bilateral loans or loans converted into grants. Where possible, debt cancellations or conversions are expected to be additional to existing ODA program lending.

(3) Multilateral creditors. Multilateral institutions have been encouraged to make further and more concessional loans and grants to sub-Saharan Africa. Virtually all of this resource transfer will be through the IDA. IBRD funds will be available for Nigeria. The IMF is to play its part through its own soft loan windows: the Structural Adjustment Facility and the Enhanced Structural Adjustment Facility.

(4) Private creditors. Nigeria, in particular, is to be encouraged to buy back or to exchange some of its commercial debt at a discount. Funds will be available to support this process through the offices of the World Bank Group.

Briefly described, this is the situation which existed in Africa at the end of 1990. In addition to the Brady and African initiatives, the years 1989–90 saw two other developments which deserve to be signalled here. First, there has been a 'parallel Brady' programme in countries such as Chile and Argentina, where debt–equity swaps have been arranged outside the rubric of the Brady initiative. Increasingly, these swaps are linked to local privatization programmes, as in Argentina where debt-equity swaps are being used to privatize the national airline, telephone companies and other public sector corporations. Second, the World Bank is urging all creditors to recognize that some countries lie beyond the terms of the Brady initiative and the Toronto Terms/Special Program for Africa. The Bank urges that care be taken to continue to reward countries such as Thailand and South Korea, which have managed their debts in a sensible manner. Attention must also be given, on a case-by-case basis, to non-African SILICs where significant signs of adjustment are apparent and are in need of suitable rewards.

Assessment

It is not yet possible to offer a full evaluation of the market-menu approach to the debt crisis. The best that we can do is to signal some positive signs associated with the new initiatives, before entering some cautionary words on the debt situation as it appeared in 1989–90.

The calendar year 1989 did mark a slowing down in the rate of increase in the total debt stocks of the developing countries. Total external debt rose by just $10 billion, from $1136.5 billion in 1988 to $1146.7 billion in 1989. In Latin America total external debt fell by $5.4 billion, an amount equivalent to a nominal rate of growth of debt in the region in 1989 of −1.3 per cent. (The SIMICs as a whole recorded a nominal rate of growth of debt in 1989 of −0.4 per cent; all other categories of debtor − SILICs, moderately indebted low income countries, moderately indebted middle income countries and other countries − recorded positive nominal rates of growth of debt stock.)

This encouraging performance should not be read uncritically. Although voluntary debt reduction operations (buybacks, swaps and exchanges) did reduce the debt stock of the developing countries by $19.9 billion, an estimated further $16.5 billion of the debt stock was reduced simply by virtue of a strengthening of the

US dollar (and associated changes in the value of dollar-measured debt stocks). Net flows on debt of $24.5 billion, together with net increases in interest arrears and rescheduled interest payments, were sufficient to offset the negative changes so that the final net change in the debt stocks of developing countries in 1989 was plus $10.2 billion. In terms of currency-adjusted rates of growth of debt stock, only Latin America and the Caribbean are reported as maintaining a negative figure (−0.7 per cent). In South Asia the currency-adjusted rate of growth of debt stocks in 1989 was as high as 8.8 per cent; in Sub-Saharan Africa it was 4.6 per cent.[7]

In terms of the Brady initiative countries specifically, there are some further positive points to report. In the three completed operations (Mexico, Philippines and Costa Rica), 'the face value of debt to commercial banks was lowered by US $9.5 billion' (World Bank 1991a, 6); the scheduled completion of the Venezuela Agreement would reduce debt stocks by a further $2 billion by the end of 1990 (World Bank estimate). Finally, 'New money totalling US $3.5 billion will be provided as a result of the operations in Mexico, the Philippines and Venezuela. Multilateral, bilateral and bridge loans of US $6.4 billion, plus use of US $1.5 billion of the countries' reserves, financed buybacks and the cancellation of principal and some interest payments in the three completed operations' (World Bank 1991a, 6).

A similar tale can be told of some African countries, at least with regard to the Toronto Terms package. Not only was the first phase of the SPA active in 21 countries in 1989, but debt forgiveness totalled $6 billion by the end of that year. As a result of this ODA debt forgiveness, 'scheduled debt service payments of these [low-income African] countries in 1990 are lower by an estimated US $100 million' (World Bank 1991a, 93). Finally, a total of more than $5 billion has been consolidated under the Toronto terms for Paris Club reschedulings. 'The cash flow savings as a result of these reschedulings amount to about US $100 million on an annual basis' (World Bank 1991a, 94).

There is, then, some progress to report. By the end of the 1980s the stock of developing countries' external debt was beginning to level out (figure 2.10), and payments of interest were beginning to reduce as a percentage of total debt service payments. In terms of debt–export ratios the story is again an encouraging one, mainly on

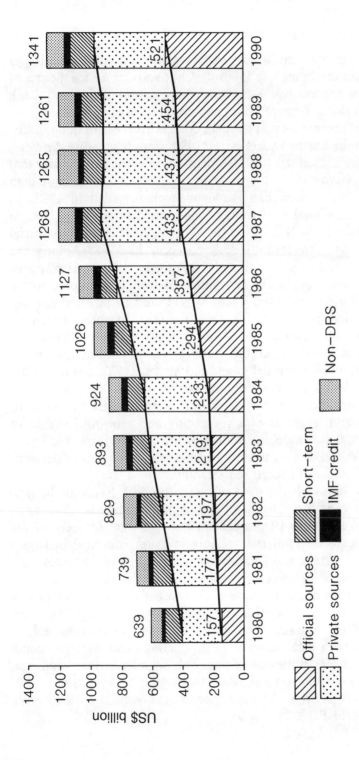

Figure 2.10 External debt of developing countries, 1980–1990
Source: World Bank (1991a)

account of developing country exports growing at more than 10 per cent per annum in the late 1980s. In Chile and Costa Rica the rate of growth of exports was especially marked, and the World Bank acted in 1989 to remove both countries from its list of SIMICs.

None of this means that the debt crisis is over, or anything like it. Although the last two years of the 1980s were relatively calm years by the standards of what had happened before, the fact remains that the total external debt of developing countries in 1990 is more than 100 per cent higher than the debt stock recorded in 1982. In addition, of a total outstanding long-term debt stock in 1990 of $1051.1 billion, $521.2 billion (or 51.3 per cent) is owed to official creditors, as compared to 36.8 per cent in 1982. Among the minutiae of debt figures, this is a significant statistic. It confirms that the private banks are becoming relatively less exposed to the problems of the indebted Third World (and to possible moratoriums on debt payments). The banking crisis has been dealt with quite effectively, even if the banks did finally have to pay a price for their 1970s activities in the form of debt write-downs, lower dividends, and discounted swaps and buybacks. To put it another way, the debt crisis has been contained in part by a process of socialization, both in terms of tax incentives in the creditor countries and in terms of a continuing and growing issuance of official credits to indebted countries since 1982. (Japanese banks are a notable exception to this rule. Their involvement in developing country markets increased steadily in the 1980s.)

Finally, although the market-menu approach has something to commend it at the margin, it should be remembered that (as of mid-1990) less than 10 per cent of the total external debt of the developing world (and then mainly in Latin America) had been touched by these new instruments and facilities. Further, the effects of some of these conversions, or reschedulings under the Toronto terms, are not always as they might appear. In the case of the Brady initiative countries, it is difficult to disentangle the direct effects of the market-menu approach from the broader effects associated with local adjustment programmes and regional economic multipliers. In the case of some sub-Saharan African countries, the impact on debt service of debt forgiveness has been quite small, 'because of the original highly concessional nature' of the forgiven loans (World Bank 1991a, 93).

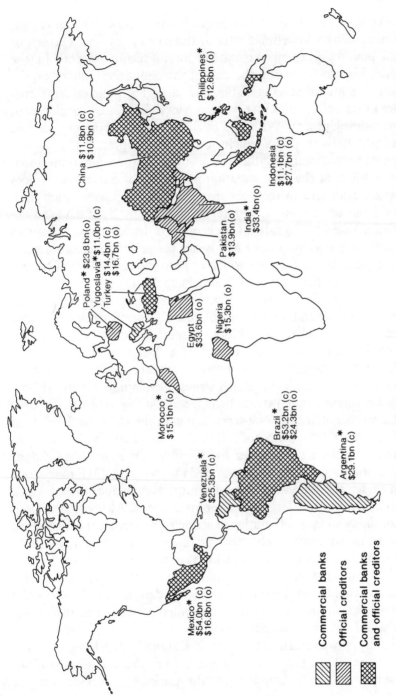

Figure 2.11 The debt crisis in 1989: non-OECD countries with a total long-term debt stock outstanding of US$10 billion or more

* Rescheduling difficulties in the 1980s. c, Owed to commercial banks; o, owed to official creditors

Source: World Bank (1991a)

Labels on map:

Philippines* $12.6bn (o)

China $11.8bn (c) $10.9bn (o)

Indonesia $11.1bn (c) $27.7bn (o)

India* $33.4bn (o)

Pakistan $13.9bn (o)

Poland* $23.8 bn (o)
Yugoslavia* $11.0bn (c)
Turkey $14.4bn (c) $16.7bn (o)

Egypt $33.6bn (o)

Nigeria $15.3bn (o)

Morocco* $15.1bn (o)

Venezuela* $25.3bn (c)

Brazil* $53.2bn (c) $24.3bn (o)

Argentina* $29.1bn (c)

Mexico* $54.0bn (c) $16.8bn (o)

Legend:

Commercial banks

Official creditors

Commercial banks and official creditors

Beyond such details, the broader picture is none too comforting. Mathematicians sometimes refer to the concept of 'gross error', or to the possibility, to mix a metaphor, of not seeing the wood for the trees. This can happen in the case of the debt crisis, where a quite reasonable attention to statistics and changing policies can sometimes blind one to the wider panorama of debt-related development or non-development. With this in mind it is sobering to note just three points by way of a conclusion to this chapter.

First, notwithstanding the market-menu approach and the apparent stability of the debt situation in 1990–1, there are still voices calling for an end to the debt management proposals which have come from the creditors. Reference has already been made to the initiative taken by President Garcia of Peru. In subsequent chapters we will have reason to refer to a wider history of sovereign default and to political movements which oppose the disciplines of 'sound finance' under the banner of 'Can't Pay, Won't Pay'.

Second, the debt crisis may yet be spreading. Although South Korea and Thailand, along with Chile and Costa Rica, are regularly cited as successful or ex-debtors (chapter 3), India must now be counted as an unsuccessful new debtor. Between 1982 and 1989 India's total external debt more than doubled to $62.509 billion, in the process making it the fourth largest developing country debtor (if not a threat to the private banks; see also figure 2.11). In 1990 India was negotiating a long-term loan with the IMF worth between five and seven billion US dollars; in 1991 an austere budget was put in place by the new Minister for Finance, Dr Manmohan Singh, and further negotiations were entered into with the IMF to secure a loan under the Extended Fund Facility. Meanwhile, the USA has become the world's largest net debtor country. Although the USA is not likely to face difficulties in servicing its debts, the manner in which this enormous debt stock is financed has potentially quite grave implications for other indebted countries (see chapter 5). Still further problems are on the horizon in the former Soviet Union and even, perchance, in Australia, where 'net debt in late 1988 was $80 billion, making it the seventh most indebted nation in the world' (Daly and Logan 1989, 35).

Finally, if we return to the main indicators of development at the end of the 1980s, the message we take from them is nothing if not alarming. In Latin America and the Caribbean average real per capita incomes were lower in 1989 than they were in 1978 and in many countries infant mortality rates remained over 40 per 1000 live births. Meanwhile, in sub-Saharan Africa, total external debt in 1989 was only fractionally less in dollar terms than the combined GNP of the region; infant mortality rates here commonly remained at over 90 per 1000 live births. These figures cannot be explained by the debt crisis alone, of course (and certainly not in the case of sub-Saharan Africa, where a number of countries have been afflicted by war and famine besides). They do remind us, however, that the debt crisis is by no means over yet: a banking crisis may have been tidied up, but a development crisis is in full swing. The question of *why* this should be the case, and what might and should be done about it, is the point of focus of the second part of this book.

Notes

1   On the similarities between the 1930s and 1980s, see Eichengreen and Portes (1989). Eichengreen and Portes (p. 40) point out that creditor governments were involved in debt negotiations in the 1930s and that 'the readjustment of defaulted interwar loans universally involved some burden sharing among debtors and creditors'. See also Díaz-Alejandro (1983), Fishlow (1985) and Maddison (1985).

2   The confidence of the international financial community was further shaken by the death in suspicious circumstances in mid-1982 of Roberto Calvi, a banker at the centre of an unfolding banking scandal involving the Ambrosiano Bank and the Vatican Bank (see Lernoux 1986, chapters 9 and 10).

3   Most of this decline can be explained by reference to the enormous expansion of world trade between 1950 and 1980 (see Grimwade 1989).

4   Wider discussions of structural adjustment programmes and IMF conditionality can be found in Haggard (1986), Streeten (1987) and Bacha (1987). On the same issues in an African context, see Parfitt and Riley (1988), Green (1989) and Muntemba (1989). On the experience of liberalization and stabilization programs in the Southern Cone of Latin America, see Corbo and de Melo (1985) and Tokman (1985).

5   On the ability of some banks to cope with debts in arrears by a strategy of securitization, see Congdon (1988). On the relationship between securitization and the emergence of a 'new international financial system', see Thrift and Leyshon (1988).

6   This chapter has called the attention of the reader to certain discrepancies which exist between and within various systems for reporting statistics on international debt. It is worth noting, also, that the production of economic statistics for many countries in the developing world is problematic to say the least. On the political context for the production and dissemination of economic statistics in sub-Saharan Africa, see Hill (1986), Watts (1989) and Yeats (1990).

7   This book does not discuss the distinctions which should be drawn between debt relief, debt reduction and reduction of the outward net transfer. These distinctions are clearly of importance to debates on the practicalities of debt crisis management: for a useful review of the issues, see Corden (1991); see also Abbott (1989).

Part II

Decoding the Debt Crisis: Discourses on Debt and Development

THREE

The Debt Crisis: a System-stability
Perspective

1 Introduction

Arguably the most optimistic set of arguments on the developing
countries' debt crisis are to be found within a system-stability
perspective. These arguments might also be accounted the most
punitive. As a first approximation to the system-stability perspective, let us define it as a set of arguments which suggest: (a) that the
international economic and financial system is inherently stable,
with open economies tending to a spontaneous order and/or to a
state of dynamic equilibrium and economic growth; (b) that for
states or international organizations to seek to intervene in open
markets on the basis of what are localized and temporary disequilibria (the so-called debt crisis) is to risk economic and political
outcomes which are third-best and worse; and (c) that responsibility for particular debt crises should be shouldered by those
creditors and debtors which have acted against the norms of
economic prudence and the imperatives of sound money.

Thus described, the system-stability perspective (SSP) may seem
more unitary than it really is. To soften the outlines of this first
approximation to a SSP, the rest of the chapter is organized as
follows. Section 2 sets out a series of general statements on
economic and political actions on which most system-stability
models draw. These general principles include ideas which relate
to: subjective preferences; notions of market competition, order
and equilibrium; the theorem of the second best; and a libertarian
philosophy of just deserts. This section also alerts the reader to

certain differences of approach within a system-stability framework, and in particular to those distinctions which mark out neo-classical, monetarist and Austrian economists. The significance of some of these distinctions will become apparent later in the chapter.

Section 3 is a bridging section which considers how the ideas set out in section 2 have become central to a counter-revolution in development theory and policy. This section shows how a long-standing developmentalism has been challenged both intellectually, under the banner of subjective preference theories and monoeconomics, and politically, in and through the higher reaches of organizations including the IMF and the World Bank.

Section 4 explores the ways in which the counter-revolution in development studies have informed a related body of work on the origins and significance of the developing countries' debt crisis. Two models are outlined here: a debt-cycle model which allies an optimistic account of temporary disequilibria in the global economy to a discussion of the 'fear of fear itself'; and a profligacy model which focuses upon issues of Pharaonism, fiscal deficits, over-valued exchange rates and capital flight. The two models are not incompatible and scholars from other traditions will accept some of their claims.

Section 5 considers how various system-stability models inform attitudes to debt crisis management. Not surprisingly, the emphasis is seen to be on structural adjustment in the short run, on getting prices right in the longer run, and on possible market write-downs of debt and debt service. Attention is also directed to the achievements of South Korea and to the moral dangers inherent in *dirigisme*, debt repudiations and non-market debt and debt service reductions. Section 6 offers a brief critique of the system-stability perspective. This is mainly in the form of questions that might be put to system-stability theorists.

2 Subjective Preferences, Invisible Hands and Spontaneous Orders

Most non-economists could offer a working definition of what constitutes 'market economics'. It might, for example, be taken to denote a belief in the power of competitive markets to bring together individuals and firms in such a way that a local problem of economic scarcity is solved efficiently from the point of view of all participants. Supply and demand schedules would be central to any such iteration of competitive market principles, as would be the institution of private property. Some non-economists might also be happy to link such a perspective within economics to broadly right-of-centre political parties and to corresponding political philosophies. The market, the freedom of the individual and the invisible hand linking individuals to each other are then seen to be closely entwined.

There is something to be said for this non-economist's perception of market economics, and certainly it catches a good deal of what will be said in this chapter by another non-economist. There is a sense in which market economics, the New Right and a system-stability perspective share a common outlook. But there are also problems with this account, as further reflection makes clear. For example, if market economics is the right label for what we have just described, where does this leave Keynesian and Marxist economics? Are they not also concerned with the market, or is it that they consider the market in different terms? Alternatively, might the definition that we have offered be too inclusive? Does it make sense to refer to a market economics in the singular when it includes such related and yet departing perspectives as neo-classical economics, monetarism and Austrian economics? Finally, in what sense should a celebration of the market be considered to be right wing or in tow to the New Right? Where does this leave the libertarians and the true conservatives?

Perhaps we can make the picture a little clearer, without trying to suggest that we can arrive at a definition which will satisfy economists and non-economists alike. When non-economists refer to market economics, what they tend to have in mind is the subjective preference paradigm. Unlike some Keynesians and most Marxists, economists in this mould begin their accounts of economic life with the subjective preferences of individual economic actors. The individual, in this perspective, is deemed to be indivisible and essential. He or she has certain taken for granted capacities and wants which, when properly understood, yield important insights into the workings of economic institutions and laws.

We can consider this proposition in more formal terms. According to Wolff and Resnick, 'three concepts taken together – individual tastes (preferences), the production function and individual resource endowments – form neo-classical theory's conceptual points of entry' (Wolff and Resnick 1987, 46). Forget, for the moment, the equation of subjective preference theories with neo-classical economics. What is being suggested here is that 'market economics' functions on the basis of an account of individual wants and productive abilities. More precisely, the direction of causality in such economic theory is from the individual to some larger or wider notion of economic institutions and activity. The individual, in the words of Milton and Rose Friedman, is 'free to choose' (Friedman and Friedman 1980). Because individuals have different talents and desires, some will prefer to work less and to have more leisure time, while others will choose a different combination of these two utilities. By the same token, some individuals have the talents and the disposition to work for high rewards, while others lack these said talents and dispositions (and rewards).

Economic activity takes place on the basis of these subjective preferences and endowments. Because we know something about human nature, however, we can be confident that a seemingly random process of individual decision-making will disclose certain patterns and law-like regularities. For example, we can assume that all individuals will prefer 'more than less of any good or service. This is sometimes referred to as the assumption of nonsatiation' (Wolff and Resnick 1987, 51). Individuals will also have to rank their preferences. This is because the range of possible choices before them will far exceed any means of purchase they have at their disposal. This ranking of preferences in turn will involve a willingness to substitute one good or service for another, usually on the basis of a diminishing marginal rate of substitutability. Finally, all individuals can be assumed to be 'rationally motivated, choice-making machines; no matter what differences [of culture, gender etc.] may separate [them]' (Wolff and Resnick 1987, 51).

When these patterned subjective preferences have been derived, the next question for 'market economics' concerns the possibility of order or equilibrium within an economy. How can the economic abilities, resources and preferences of all economic agents be coordinated such that certain desirable properties of economic and political systems (efficiency, liberty, transparency etc.) are maximized? The answer to this question is presumed to lie with the market. Since the time of Adam Smith it has been a common assumption in economics that 'initial private actions can have beneficial public effects that were not intended by the actors' (Vaughn 1989, 169). Whether they will or not depends upon the workings of the invisible hand and the market conditions which make its workings possible. Subjective preference theorists are inclined to emphasize the virtues of a competitive market economy. On the assumption that market prices are dictated by the interlocking demand and supply schedules of individuals who are not able to dictate those prices, subjective preference theorists suggest that a market equilibrium will arise wherein total economic production is unintentionally maximized.

The foregoing is doubtless too simple, yet some important points have already been made. The 'market' is not the basic building block of what is often called market economics; that distinction falls rather to the individual and to his or her resources and subjective preferences. Competitive markets then emerge as the most effective means for the coordination of diverse economic activities and exchanges between freely transacting agents. By the same token, the market is not studied or celebrated for its own sake, any more than it is in so-called non-market economics. The market is feted as a means to an end.

What these ends might be varies according to different shades of opinion within the subjective preference paradigm. In the case of neo-classical economics, markets are lauded for their capacity to secure welfare-efficient outcomes. Neo-classical economics derived from what is now called classical political economy in the second half of the nineteenth century. In the work of theorists including Jevons and Walras, the political and institutional frameworks for economic actions were progressively played down until a series of economic principles could be derived on the basis of various abstract propositions and assumptions. Some of these assumptions had to do with a state of perfect competition. In neo-classical economics, it is common to make economic predictions and claims on the basis of a model which first assumes: (a) that there is no divergence between private and social costs anywhere in the economy; (b) that there are no external economies or diseconomies of scale; and (c) that the process of innovation and growth is strictly autonomous (after Lipsey 1975, 310). On the basis of these and other assumptions, neo-classical economists have been able to suggest that 'there exists a set of prices in all markets under which everyone will be satisfied to the extent that nobody wishes to exchange further at those prices' (Cole et al. 1983, 96). This is an extraordinary conclusion for it implies not only that competitive markets are efficient in a state of general equilibrium, but also that these markets are Pareto optimal (in the sense that nobody can become better off without making others worse off). Except in certain restricted cases, any interference with such a competitive market must be counter-productive.

Not all subjective preference theorists defend the market in these terms. It is significant that a second strand of subjective preference theories – the Austrian school founded by Menger and continued by Mises and Hayek among others – is less enamoured of abstract mathematical modelling and of the general equilibrium solutions which are thereby produced. The Austrians suggest that markets are to be preferred as social and economic institutions because market prices are the best available means of signalling economic information, which is scarce, uncertain and decentralized in origin. Open and competitive markets are also a means to preserve the liberty of the individual. Markets perform this function by removing from monopolies the power of arbitary resource use. They also embody the quality of non-intentionality that Adam Smith so ably claimed for them. For most Austrian economists, real markets cannot reasonably be reduced to the tenets of perfect competition theory and general equilibrium analysis. Markets are rather defined by a 'rough and tumble process of market agitation kept in motion by complete freedom for competitive entrepreneurial entry' (Garrison and Kirzner 1989, 123). In other words, markets are not about end-states or intentions at all, but are about a continuing state of change wherein new 'possibilities and preferences [are discovered] that no one had realized hitherto' (Garrison and Kirzner 1989, 123). By the same token, markets are the results of human action, but not of human design (a favourite aphorism of Hayek's). The fundamental institutions of society owe their existence to no identifiable creator or necessary tendency to equilibrium; they simply display what Hayek called a *spontaneous order* (Hayek 1960, 160). Money, like language, is just such an instance of a spontaneous order. Money is an institution which came into being as a 'result of a long sequence of actions on the part of a multitude of traders, none of whom *intended* to create the institution of money' (Garrison and Kirzner 1989, 121; emphasis in the original).

It follows from this that Austrian economists are less inclined than neo-classical economists to ask 'what institutions would be necessary to make the invisible hand work perfectly'; Austrians rather pose 'the question, what are the economic reasons why existing market institutions emerged and what unperceived purposes do they serve' (Vaughn 1989, 172). Notwithstanding these differences, however, Austrian and neo-classical economists share a good deal of common ground when it comes to questions of order and equilibrium. Certainly, they are each at some remove from the Keynesians and the Marxists. This much is evident as soon as we consider the question of disorder or disequilibrium.

In the context of post-war economic theory, it is not unreasonable to suggest that the problem of disorder has been of paramount importance. The resurgence of subjective preference theories is due in no small part to the willingness of neo-classical and Austrian economists to challenge the wisdom of a prevailing Keynesianism. Hayek's work, in particular, might be considered as a forty-year struggle with Keynes's intended and unintended legacy. For Hayek, as for many others, Keynesianism is invidious because it presents a threat to the open society. It does this by virtue of its willingness to think in terms of outcomes which could be willed or intended by institutions (as when the state commits itself to an incomes policy or to a goal of full employment). Most subjective preference theorists resist this presumption. They point out that such intended outcomes can only be secured in the short run, and then only with disturbing implications for the long run health of an open economy. More especially, they have tended to point out (and this is important for our story) that most distortions to the market have come by means of monetary infusions and other financial disturbances.

The role played by money in economic booms and busts reveals further variations within the subjective preference framework. Perhaps the best known modern theorist of money is Milton Friedman, but his version of closed-economy monetarism is not shared by all subjective preference theorists (Friedman 1960, 1969). Friedman's work adapts the quantity theory of money to argue that a destabilizing inflation is always and everywhere the result of too much money being placed in circulation. Hayek accepts that the equation of exchange ($MV = PY$) is true by definition (the quantity of money multiplied by the number of times this money is spent in a given year must equal nominal income), but he is sceptical of the weight which monetarists place on a relatively stable velocity of circulation of money. Hayek would prefer to put the emphasis upon the destabilizing impact of 'money injection effects', regardless of the value of money as measured by the general level of prices. Where Hayek would agree with Friedman, and with constitutional economists like James Buchanan (the intellectual 'father' of the community charge/poll tax in the UK), is in his presumption that governments are inclined to buy short-run Keynesian goals by means of monetary infusions which can only be inflationary and damaging over the longer term. It is this presumption which is of interest to us.

Hayek and Friedman would also agree on certain political and moral claims which are compatible with versions of subjective preference economics. We can end this brief survey of general ideas by noting just four of these claims. First, it is a general presumption of subjective preference theories that actions which interefere with the workings of competitive markets must be sub-optimal in terms of efficiency. As Deepak Lal puts it: 'no general rule of second best welfare economics permits the deduction that, in a necessarily imperfect market economy [he refers to the developing world], particular *dirigiste* policies will increase welfare'. Indeed, *dirigiste* policies have often 'led to outcomes which, by the canons of second-best welfare economics may have been even worse than *laissez faire*' (Lal 1983a, 10). Related to this is a second presumption: that proposals for planned or non-market economies can only sensibly be based on the very market price signals which are intentionally displaced. (This idea is relevant to debates on 'internal markets' – as in public health care delivery systems).

A third claim can be described as a just deserts argument. For most subjective preference theorists, and certainly for libertarians like Nozick and Hayek (Nozick 1974; Hayek 1960), the idea that governments should be involved in the pursuit of greater equality is economically ill-founded and politically dangerous. The proposition is based on a prior assumption that inequality is a bad thing, that it is a euphemism for unfairness, and that it can be rectified by state-led transfers of income and wealth. Subjective preference theorists are minded to dismiss such claims: inequality, far from being a bad thing, is entirely natural in a free society. Inequality is the logical result of society comprising a diverse set of economic agents, each of whom brings to market different tastes and abilities, and each of whom deserves a different reward.[1] Unless hard work and innate ability are to go unrewarded, inequality is the norm and equality is something which can only be imposed by those who would claim to know what others might want and deserve better than the individuals concerned. As Kenneth Minogue puts it: 'Inequality is notoriously the common condition – one might well say "the natural condition" – of mankind. . . .My disagreement with egalitarianism is thus both particular and general. I oppose not only the plan of equality, but any kind of plan at all' (Minogue 1990, 100). The same logic can be extended to inequalities between countries.

Finally, there is the question of market rules and the rule of law. Subjective preference theorists are not generally proponents of *laissez faire*, nor are they always politically conservative (in the sense of resisting change). Most subjective preference theorists prefer to acknowledge that the state does have a role to play in guaranteeing the conditions of existence of markets and liberty. This will include, for example, the provision of certain public goods (including defence), and the provision of a system of law and order which is prospective and rule based, while acting to secure well defined property rights. The state might also sponsor certain improvements in local communications systems on the grounds that this reduces transaction costs, but this is a moot point. Whether or not the state should be involved in the issuing of money is similarly a matter of dispute among subjective preference theorists. Most monetarists and constitutional economists accept that the state should issue money, but they insist that states should do so in a manner which corresponds to certain rules of political non-interference. In other words, central banks should have the power to resist political temptations to buy votes. They would do this by issuing money in accordance with sensible and previously published monetary targets. Hayek, characteristically, goes a stage further. In the 1970s he arrived at the view that monetary stability – neutral money – could only be achieved by turning over to private enterprise the capacity to issue money (Hayek 1976). In effect, individuals would be free to choose one currency or form of money over another 'on the basis of the issuer's demonstrated ability to achieve purchasing power stability for that currency' (Garrison and Kirzner 1989, 129). Competition would act to minimize the potential for 'dis-coordination' inherent in money, while allowing 'market participants to take the fullest advantage of the remaining elements of the spontaneous order' (Garrison and Kirzner 1989, 129).

3   The Counter-revolution in Development Theory and Policy

When development studies emerged in the 1950s and 1960s its primary attachments were not to the claims of a subjective preference paradigm. Development economics rather attached itself to two main sets of propositions: (a) to the view that economic development should be sponsored as an industry-based process of structural transformation, aided and abetted by inflows of resources from developed countries; and (b) that markets within developing countries, and between developed and developing countries, were so prone to failure and asymmetry that government actions in support of protectionism and/or infant industries were necessary and legitimate.

In more recent years this developmentalist perspective has been attacked by the left and by an emerging counter-revolution in development theory and policy. From the latter quarter comes the charge that *dirigisme* for development has failed an inward-looking Third World, with the theorem of the second best being perverted to sponsor policies which can only hinder local processes of economic development. The charge has also been floated that mainstream development economics is corrupt and impoverishing.

Lord Peter Bauer and Deepak Lal have been especially forceful in blaming a development duoeconomics – or the view that markets work in developed countries but not in developing countries (Hirschman 1981) – for the landscapes of poverty and disrepair which so often typify parts of the developing world (Bauer 1972, 1981, 1984; Lal 1983a).

Gottfried Haberler is another who has added to this critique of duoeconomics as *the* bastard child of Keynesianism and social democracy. He has joined with Bauer and Lal in deriding the view that development can be sponsored quickly by massive injections of capital. He also rejects the view that it is the responsibility of the richer countries to provide alms for the poorer countries (this takes us back to the just deserts argument). Finally, Haberler suggests that a willingness to create rent-seeking societies is implicit in duoeconomics, and is based upon the mistaken view that mature capitalism contains within it a necessary tendency to secular stagnation. Although the events of the 1930s – continual deflationary pressures, declining terms of trade for commodity producers, competing economic nationalisms – encouraged Prebisch and others (Singer and Nurske come to mind) to believe that import-substitution industrialization was the key to development, Haberler insists that 'the theory of secular stagnation is a gross misinterpretation of the Great Depression' (Haberler 1987, 56). He continues:

Actually, the Depression of the 1930s would never have been so severe and lasted so long if the Federal Reserve had not by horrendous policy mistakes of omission and commission caused or permitted the basic money supply to contract by 30 per cent. One need not be an extreme monetarist to recognise that such a contraction of the money supply must have catastrophic consequences. In other words the Great Depression was not a crisis of capitalism, as Prebisch says, but was a crisis of largely anticapitalist government policy, the consequences of horrendous policy mistakes. (Haberler 1987, 56: note well this idea of policy mistakes, especially in the arena of money; we will return to it.)

In this manner, Haberler joins with other counter-revolutionary theorists in arguing against a once orthodox development economics. Just as Schultz has lauded the market rationality of optimizing peasants (Schultz 1964, 1987), so Haberler, Little and Balassa (and others too numerous to mention) have made plain their faith in monoeconomics and in outward-looking development strategies (Haberler 1959, 1985; Little 1982; Balassa 1981). According to this view, markets exist in the Third World along with a market mentality. The proper function of government is to let the market mechanism flourish (as it has done, allegedly, in countries like South Korea and Taiwan). As the *Economist* magazine put it in 1989: 'what the world's poor countries need most is less government' (*Economist* 1989, 56).

In the 1980s the counter-revolution in development theory was sufficiently well formed that its perspectives were taken seriously by policy-makers in institutions including the World Bank (where Anne Krueger was influential) and the IMF. As the World Bank moved away from a discourse which emphasized basic needs and redistribution with growth (the McNamara–Chenery years of the 1970s), it found a new vision in keeping with many of the claims made by subjective preference theorists. This new vision was evident in the Bank's prescriptions for *Accelerated Development in Sub-Saharan Africa* (the so-called Berg Report on the importance of agricultural pricing policy and the perils of overvalued exchange rates: World Bank 1981), and in its support for structural adjustment programmes. More directly, the new vision was announced at the beginning of the World Bank's *World Development Report* for 1985. According to this report, 'the record of development and the growing store of empirical research have heightened recognition of the importance of markets and incentives – and of the limits of government intervention and central planning. The new vision of growth is that markets and incentives can work in developing countries. But they are filtered through government policies which, if inappropriate, can reduce or even negate the possible benefits' (World Bank 1985a, 1).

4　The Debt Crisis: Two System-stability Models

It is worth pointing out that the World Bank is not a monolithic institution and that not all of its senior staff would have subscribed to the views just outlined, even in the mid-1980s. The World Bank is an institution which is subjected to many pressures and it is not a little pragmatic in some of its policies (something which the counter-revolution cannot be accused of). Moreover, there are signs that the World Bank is now rethinking some of its attitudes to markets and states. In the *World Development Report* for 1991, Bank officers acknowledge that 'markets sometimes prove inadequate or fail altogether' (World Bank 1991b, 1). Although 'competitive markets are the best way yet found for efficiently organizing the production and distribution of goods and services' (ibid.), it is unwise to be dogmatic about this or to oppose intervention to *laissez faire* – 'a popular dichotomy, but a false one' (ibid.). On page 26 of the report favourable attention is even given to a Scandinavian model of development.

These points made, it is still significant that the rise of the counter-revolution in development theory and policy coincided with those years in which inflation and debt became major issues requiring intellectual and practical consideration (Stewart 1984). Given this correspondence, it is not surprising that the debt crisis received prominent attention from theorists and practitioners who might be seen as proponents of the counter-revolution. Scholars including Beenstock and Lal, Bauer, Vaubel and Sjaastad have advanced a system-stability perspective on the so-called debt crisis, which moves between two main sets of propositions: first, an early debt and development-cycle model which discounts the dangers inherent in the 'debt crisis'; and, second, a profligacy model which links external debts to budget deficits. We can consider them in turn.

Debt and Development Cycles

According to perhaps the most prominent of debt cycle theorists, Michael Beenstock, 'there is no aggregative debt problem and there is no fundamental threat to world monetary stability. . . .There is nothing to fear but fear itself' (Beenstock 1984, 224–5).

To support this claim Beenstock puts forward four theses on the so-called international debt and banking crisis. He first reminds his readers that debt is not intrinsically damaging; indeed the capacity to accept credit must imply a degree of creditworthiness and a measure of economic growth. From this insight Beenstock develops a 'development cycle theory of indebtedness by nations' (Beenstock 1984, 242). For Beenstock, nations, like individuals, tend to acquire debts during the youthful phase of their development cycles (see figure 3.1): 'Just as private individuals are highly leveraged during the early part of the life cycle so developing countries tend to become highly leveraged during the early stages of the development cycle' (ibid.). This is as true in Latin America today as it was true in the white settler colonies 70 years ago. Indeed, 'viewed in these historical terms today's developing countries have surprisingly small debts. . .[and] are under-geared in comparison with some generations of developing countries' (Beenstock 1984, 231; and see table 3.1). Beenstock concludes that, 'just as individuals repay their debts as they mature so do nations reduce their indebtedness [and their trade surpluses] as they reach the maturer phases of the development cycle' (ibid., 242).

Table 3.1 Comparative long-term debt–export profiles

Long-term debt of non-oil developing countries		
	1973	1982
Total external debt (US$ billions)	97	505
As percentage of output	20	30
As percentage of exports	90	110
Foreign debt as a percentage of exports in 1913		
Canada	860	
South Africa	630	
Latin America	520	
Australasia	480	
Russia	480	
Japan	230	
China	220	

Source: Lal (1983b)

If indebtedness 'is largely an equilibrium phenomenon explicable in terms of the development cycle theory' (ibid. 228), then so too is the appearance of an aggregative debt crisis the result of a temporary shortfall in country liquidity. This is the substance of Beenstock's second point. Beenstock believes that the world economy is in a state of transition wherein 'the greatest challenge to the world economic order involves coming to terms with the inexorable spread of economic development over the next hundred years' (Beenstock 1984, 226). Beenstock further suggests that the balance

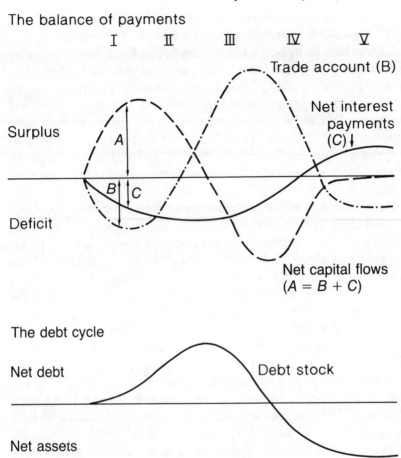

Figure 3.1   Balance of payments flows and debt stock during the debt cycle
Source: World Bank (1985b, 47)

of world economic power began to shift in favour of the developing countries in the 1960s. Against this background the 'debt crisis' must be counted as an unfortunate interruption to the onrushing march of development. Beenstock insists that the origins of the debt crisis must not be looked for in terms of a credit-rationing model, or via a profligacy thesis, or even as a result of the exogenous shock of oil price rises; instead a temporary crisis of liquidity was 'induced by the counter-inflationary policies of OECD countries especially since the late-1970s' (ibid. 257).

This brings us to Beenstock's third point. Because the 'liquidity crisis is temporary in nature. . .[it] may disappear as suddenly as it arose' (Beenstock 1984, 236). In more detail: 'The prospects are that real commodity prices will increase and real interest rates will abate as OECD economies recover once inflation has been squeezed out. The debt "crisis" will then disappear as suddenly as it appeared and growth in the Third World and transition in the world system will resume their *normal* course' (ibid. 259; emphasis added).

Beenstock does not tell us when these events will take place, and it is not clear why he looks to an OECD locomotive effect when his transition theory suggests that the balance of world economic power is shifting southwards. Instead, Beenstock's fourth point concerns the lack of threat which this so-called debt crisis presents to a stable international economic order. Beenstock rejects the claim that the failure of a money-centre bank will provoke 'an epidemic of closures. . .because of the logic of the inter-bank market' (ibid. 259). Beenstock argues that at the end of 1981 the 'banks had about 15 per cent of their assets invested in developing countries. This does not amount to an extensive degree of exposure' (ibid. 230). Even if a bank does take a loss which exceeds its equity capital, it does not follow that the bank must close. Beenstock concludes that: 'If Bank X is fundamentally healthy and the loss is unfortuitous the capital market will regard bygones as bygones and will generate the necessary equity capital to keep the bank in business. . . .One or even a few rotten apples do not spoil the barrel provided the capital market is rational' (Beenstock 1984, 259–60).

Beenstock's essay on the debt crisis is optimistic even by the standards of the system-stability perspective. It is worth setting out at length for its clarity in revealing certain assumptions and claims common to a subjective preference framework. Beenstock's optimism – his apparent *laissez faire* – is predicted on the view that markets are stable and rational, and that bad debts are no more than temporary disturbances to an underlying economic equilibrium. His 'real world' is a world of 'normal courses' and fundamentally orderly relationships. Deepak Lal takes a similar line. He joins with Beenstock in dismissing those Jeremiahs who warn blithely of high levels of outstanding debt in the Third World or of high debt ratios. Lal maintains that 'most of these ratios are meaningless. For as long as a borrower can utilize a foreign loan productively to yield a rate of return at least equal to the real interest cost of borrowing and convert the equivalent domestic resources into foreign exchange, the foreign borrowing can pose no problem' (Lal 1983b, 18). In any case, says Lal, the debt/export ratios of most developing countries are not high by historical standards. The 'ratio of long-term debts to exports of non-oil developing countries of 1.1 in 1982 is well below the lowest ratios of 2.2 for China and Japan in 1913 and a fraction of those for Canada and South Africa. There was little talk of a debt crisis then' (ibid.).

Not everyone goes this far, however, even within the debt-cycle component of a system-stability framework. In the early to mid-1980s a more cautiously optimistic tone was struck by William Cline, a Senior Fellow of the Institute for International Economics in Washington, DC, and a prominent student of the international debt crisis. More so than Beenstock and Lal, Cline is prepared to acknowledge a serious threat to the international banking system. He is also less easily categorized as a 'true believer': Cline takes a more pragmatic line on the debt crisis than does a counter-revolutionary. Cline is of interest, none the less, because his work provided a powerful exposition of the view that the debt crisis consists in a temporary (and cyclical) crisis of liquidity and not in a secular crisis of global solvency. In a most valuable exercise Cline aimed to discern 'whether the major debtor countries are illiquid or insolvent: whether their obligations should be viewed as largely sound debt or bad debt' (Cline 1984, 39). Cline added: 'If they are merely illiquid, additional lending is appropriate to tide them over short-term difficulties. If they are insolvent, it may be more appropriate to recognise their debt as bad debt and to attempt to salvage at least some portion of the debt while accepting some loss on face value' (Cline 1984, 39).

To examine this question – to see whether the system was stable or unstable – Cline set up a projection model which would chart the evolution of external debt totals and current account deficits for the 19 countries with the largest outstanding debts in the years 1982–6 (the model was later expanded and its coverage extended to 1987). The model has five parameters (see table 3.2):

1   *The rate of growth of the industrial countries*: Cline signals the importance of this variable by setting five alternative combinations of growth rates. According to Cline's worst-case assumption, the industrial economies grow at only 1.5 per cent per annum from 1984 and are stagnant in 1982 and 1983. The most optimistic scenario, E, assumes a post-recession annual rate of growth of 3.5 per cent. Significantly, Cline chooses scenario D – 3 per cent growth per annum after 1984 – as his base-case assumption. Cline thereby signals his faith in a spontaneous world recovery.

2   *The price of oil*: in the base-case estimate the price of oil fluctuates between $34 and $29 per barrel. An alternative scenario suggests a modest decline in oil prices.

3   *The rate of interest*: LIBOR is assumed to fall sharply in the base-case model, but stays more or less constant in a rival scenario.

Table 3.2   The 'Cline model': alternative global economic parameters

	1982	1983	1984	1985	1986
Industrial country growth (%)					
A	0.0	1.0	1.5	1.5	1.5
B	0.0	1.5	2.0	2.0	2.0
C	0.0	1.5	2.5	2.5	2.5
D[a]	0.0	1.5	3.0	3.0	3.0
E	0.0	2.0	3.5	3.5	3.5
Oil price ($/barrel)					
A[a]	34.0	29.0	29.0	29.0	34.0
B	34.0	20.0	20.0	25.0	25.0
LIBOR (%)					
A[a]	15.0	10.0	9.0	8.0	8.0
B	15.0	11.0	13.0	15.0	15.0
Dollar (index)					
A[a]	1.00	1.05	1.15	1.15	1.15
B	1.00	1.00	1.05	1.05	1.05
Inflation (%)	4.0	5.0	5.0	5.0	5.0

[a]Base-case assumption.
Source: Cline (1984, 45)

4 *The price of dollars*: both scenarios assume a slight devaluation of the US dollar.
5 *The rate of global inflation.*

On the basis of a simulation model which combined various of these parameters in different states, Cline was able to produce debt and balance of payments projections for each of the 19 countries (including the 11 largest debtors), for the 19 countries together, for a set of oil-importing countries and for a set of oil-exporting countries (see tables 3.3 and 3.4). The results of this exercise were broadly comforting. Cline noted that 'under the base-case for growth of the world economy ($1\frac{10}{2}\%$ in 1983, 3% annually in 1984–1986), the severity of the debt problem recedes substantially' (Cline 1984, 46). Cline did not take this to be a recipe for inaction, however, nor did he exhibit an unquestioning faith in an equilibrating market. Cline emphasized that the 3 per cent rate of growth was a threshold figure: 'if growth is $2\frac{10}{2}\%$ or below, the situation remains little improved or deteriorates' (ibid. 46). Moreover, to achieve a healthy rate of global recovery, Cline thought that it might be necessary for the leading industrial countries to pursue a new mixture of fiscal and monetary policies. In his own words: 'The industrial countries would do far better to choose policy mixes that have lower rather than higher interest rates (looser monetary policy and tighter fiscal policy rather than the reverse), for a given result in terms of real growth' (ibid.). Such a statement puts Professor Cline at some remove from the more committed counter-revolutionaries; it also confirms that a good deal of worthwhile work is done by individuals and groups not easily identified with one paradigm of economic thought.

Cline's caution was further evident in his analysis of specific country (and country-group) debt scenarios. With respect to the major Latin American debtors, Cline remained confident. His projections suggested that the total stock of external debt in Latin America would stabilize in the years 1985 and 1986 and would decline thereafter. (Remember, this is assumed to be in the absence of non-market solutions to the debt crisis.) The strong export performance of Mexico and Brazil was especially commended for causing this turn-around. Elsewhere, the future was thought to be less rosy. In the oil-exporting countries, especially, and even on the

Table 3.3   The 'Cline model': current account and debt projections –
11 major debtors, 1982–1986

		1982	1983	1984	1985	1986
Brazil	CA	−14 000	−7131	4729	−1041	−647
	D	88 200	93 060	95 843	94 231	92 347
	NDX	3816	3463	2648	2244	1965
Mexico	CA	−4254	−2321	−5899	−6970	−6005
	D	82 000	82 619	87 573	92 877	96 957
	NDX	2727	2817	2582	2526	2316
Argentina	CA	−2400	−2476	−825	257	996
	D	38 000	39 752	39 583	38 175	35 898
	NDX	3720	3383	2572	2146	1796
Korea	CA	−2219	−1720	−334	628	664
	D	35 800	37 826	38 860	38 599	38 524
	NDX	1060	0999	0827	0725	0635
Venezuela	CA	−2200	−4363	−9021	−10 681	−10 615
	D	31 285	35 537	45 075	55 781	66 473
	NDX	1042	1438	1910	2474	2639
Philippines	CA	−3500	−4110	−3779	−3581	−3946
	D	22 400	26 119	29 614	32 796	36 334
	NDX	2891	3080	2742	2612	2607
Indonesia	CA	−6600	−4568	−7318	−9770	−10 160
	D	21 000	25 844	33 611	43 623	53 948
	NDX	0895	1159	1379	1725	1900
Israel	CA	−5100	−6815	−7359	−8593	−9890
	D	20 400	27 487	35 433	44 322	54 535
	NDX	1921	2398	2504	2828	3187
Turkey	CA	−1100	−1375	−577	36	−127
	D	19 000	20 329	20 989	20 935	21 163
	NDX	2406	2334	1921	1677	1507
Yugoslavia	CA	−464	−551	830	1350	1790
	D	18 477	19 359	19 048	18 077	16 698
	NDX	1136	1041	0797	0653	0522
Chile	CA	−2540	−3911	−3479	−3290	−3865
	D	18 000	21 730	24 872	27 723	31 159
	NDX	3003	3096	2634	2551	2594

CA = current account (US$ million); D = total debt (US$ million); NDX = net debt
(deducting reserves) relative to exports of goods and services (ratio, base-case).
Source: Cline (1984, 50–1)

Table 3.4 The 'Cline model': projections of balance of payments and debt, base-case, 1982–1986 (US$ million)

	1982	1983	1984	1985	1986
Oil importers					
Exports	110 536	125 243	158 805	179 936	199 758
Imports	−125 552	−135 360	−159 308	−174 566	−194 848
(Oil)	−34 499	−29 426	−29 426	−29 426	−34 499
Interest	−29 464	−29 256	−30 058	−29 591	−30 187
Current account	−35 451	−30 890	−20 207	−12 564	−12 626
Debt	299 377	327 595	346 638	355 816	365 535
Net debt/exports (ratio)	1.94	1.88	1.55	1.40	1.28
Debt service/exports (ratio)	0.39	0.38	0.32	0.28	0.26
Oil exporters					
Exports	76 300	69 783	74 836	78 072	89 813
(Oil)	59 140	50 443	50 443	50 433	59 140
Imports	−64 756	−66 835	−84 013	−92 747	−101 070
Interest	−15 423	−16 520	−16 886	−18 305	−21 284
Current account	−20 989	−19 711	−33 973	−40 933	−40 674
Debt	184 778	201 558	234 702	272 769	310 119
Net debt/exports (ratio)	1.77	2.04	2.13	2.35	2.36
Debt service/exports (ratio)	0.34	0.40	0.38	0.40	0.41
Total, 19 debtors					
Exports	186 836	195 026	233 642	258 008	289 571
Imports	−190 308	−202 195	−243 321	−267 314	−295 918
Interest	−44 887	−45 775	−46 944	−47 896	−51 471
Current account	−56 440	−50 602	−54 181	−53 498	−53 300
Debt	484 155	529 153	581 340	628 585	675 654
Net debt/exports (ratio)	1.87	1.94	1.74	1.70	1.62
Debt service/exports (ratio)	0.37	0.38	0.34	0.32	0.30

Source: Cline (1984, 47)

base-case assumption, debt totals were predicted to rise sharply. Were oil prices to drop further, and in line with assumption B, Cline believed that some oil-exporting debtor countries could face debt-servicing difficulties.

These local difficulties do not detract from the general optimism of Cline's study. Although Cline pays more attention to the *possibility* of system failure than does the true believer, he remains convinced that the debt crisis is a temporary crisis of illiquidity. As Cline puts it: 'because the chances are good for adequate global recovery the debt will indeed be manageable and therefore it will be counter-productive to adopt, out of unnecessary panic, sweeping debt reform measures that might have adverse effects of their own' (Cline 1984, 69). Cline's main sop to the Jeremiahs is his admission that Latin America may face a 'lost decade of growth and development' (ibid. 196). Otherwise, his advice is to sit tight for recovery and to prepare only 'contingency plans' to smooth through a period of local adjustments to indebted development (see also Clausen 1983).

The Profligacy Thesis

By 1985–6 it was apparent that the optimistic assumptions which were built into most debt-cycle models were not being realized in the real world of the developing countries' debt crisis. We have noted this already in chapter 2, and in chapter 4 we will comment in detail on the work of William Cline and the associated models and predictions put forward by the World Bank. (In the depths of the recession in 1983, the World Bank's *World Development Report* felt able to conclude 'that most developing countries should be able to regain their growth momentum [even though] to do so will require a more favourable world environment, coupled with significant efforts by developing countries to make better use of their resources' (World Bank 1983b, 27). In 1984 reference was made to economic growth reviving and to attention shifting 'to the prospects for sustaining the recovery' (World Bank 1984b, 34), a theme repeated in the *Report* for 1985 which begins with the blunt declaration that 'The economic turbulence of the past few years has subsided' (World Bank 1985b, 11). In 1986 the High Case projection forecast a 5.9 per cent rate of growth in the world economy, while in 1987, despite references to a modest and weak expansion of said economy (World Bank 1987b, ii), the central message was 'for each country [to] improve the conditions for its own interaction with the rest of the world in order to benefit from improved global economic conditions' (World Bank 1987b, 35). Even in 1988 and 1989, with clear evidence of slowing economic growth and continuing high debt burdens, the Bank's forecasts remained 'cautiously optimistic' (World Bank 1988b, 36). The system was nothing if not stable, with good tidings being just around the corner.)

Against this background, some proponents of a system-stability perspective began to shift their ground. Although a willingness to blame the indebted nations for their plight had not been absent in the years 1982–5, this view was not a dominant theme. Debt-cycle theorists preferred to discount the size of external debt stocks in the Third World (which were said to be low in historical terms), and Beenstock noted that it was precisely the most indebted countries which had grown most rapidly in the late 1970s (see table 3.5). As we have said before, the so-called debt crisis was treated by most debt-cycle theorists as the result of a temporary crisis of liquidity brought about by an OECD-inspired recession. Once inflation was bought under control, the debt crisis would disappear.

In the mid-1980s this view was quietly retired. As system-stability theorists came to terms with the longevity of the debt

Table 3.5 Third World debt and development, 1970–1980

| Country | Average annual growth rates (%) | | Debt service[a] as a percentage of | | | |
| | | | GNP | | Exports of goods and services | |
	Output 1970/80	Exports 1970/80	1970	1980	1970	1980
Mexico	5.2	13.4	2.1	4.9	24.1	31.9
Brazil	8.4	7.5	0.9	3.4	18.9	22.9
Argentina	2.2	9.3	1.9	1.4	21.5	16.6
Venezuela	5.0	−6.7	0.7	4.9	2.9	13.2
S. Korea	9.5	23.0	3.1	4.9	19.4	12.2
Chad	−0.2	−4.0	1.0	3.1	3.9	n.a.
Niger	2.7	12.8	0.6	2.2	3.8	2.3
El Salvador	4.1	1.5	0.9	1.2	3.6	3.5
Ghana	−0.1	−8.4	1.1	0.6	5.2	6.0
Ethiopia	2.0	−1.7	1.2	1.1	11.4	7.6

[a]Debt service is the sum of interest payments and repayments of principal on external public and publicly guaranteed medium and long-term debt.
Source: World Bank (1982b, tables 2, 8 and 13)

crisis, attention was turned to its differential regional incidence and to the debtor countries themselves. A case-by-case approach was now re-established at the level of theory (as it was already in debt management practice). In particular, attention was focused on the problems inherent in a 'globalist' interpretation of the debt crisis. Further attention was given to the problematic nature of fiscal and trade policies in problem debtor countries.

Consider, first, the critique of globalism. Given the nature of the counter-revolution in development theory and policy, there is an extreme reluctance in this quarter to acknowledge the existence of a Third World. Peter Bauer has made a reputation for himself by insisting that 'The Third World is merely a collection of countries whose governments, with occasional odd exceptions, demand and receive official aid from the West. This is the only bond joining its diverse and often antagonistic and warring constituents, which have come to be lumped together since the late 1940s as the underdeveloped world, the less developed world, the non-aligned world, the developing world, the Third World or the South' (Bauer 1991, 41).

A similar attitude can be detected within a system-stability perspective on the debt crisis. In the mid-1980s system-stability theorists were concerned to challenge the view that the problems of the indebted countries were caused mainly by exogenous shocks. In this effort they were supported by the work of Peter Nunnenkamp, an economist whose work defies simple classification. On the basis of a detailed investigation of the regional economic effects of the 1973 oil price rises, Nunnenkamp concluded that there is not a 'strong correlation between the balance of payments impacts of negative external shocks in the form of terms-of-trade losses and the emergence of debt problems' (Nunnenkamp 1986, 59). Indeed, 'for some of the countries most severely hit by worsening terms-of-trade, the external debt situation did not cause major trouble; this was particularly to be observed in Asia. On the other hand, for Latin American and African countries which proved to be major problem borrowers, the terms-of-trade effect remained comparatively weak or even non-existent' (Nunnenkamp 1986, 59–61).

Nunnenkamp was not alone in this judgement. The World Bank, in 1990, was keen to emphasize that 'A majority of developing countries have been able to service their external debt in the 1980s, and their development progress in this decade has not been much different from that in earlier periods' (World Bank 1990a, 7–8). The World Bank accounted for this seeming success in terms of the willingness of such countries to adjust their economies to the global economy, and to make proper use of the external funds which they had previously acquired. Although South Korea is not mentioned in this passage, it is clear that South Korea falls into the group of countries just mentioned. Certainly, it is a common point of reference for most system-stability theorists. Not without reason, it has been pointed out: (a) that South Korea in 1982 had the fourth largest external debt stock of all developing countries; and (b) that 43.5 per cent of its external liabilities in 1984 were at floating rates (a higher percentage than that recorded in Argentina: Buiter and Srinivasan 1987, 414).[2] Notwithstanding (a) and (b) it is expected that South Korea will be a net creditor country in 1994 (Seth and McCauley 1987). South Korea has been hailed as a free market success story and/or as a country which has prospered by virtue of being outward-looking (the two claims should not be conflated: see chapter 4).

The success of some indebted East Asian NICs is also central to the profligacy thesis. Having established that global economic conditions cannot reasonably be held responsible for the plight of most countries in debt servicing difficulties, the suggestion then is that these difficulties must be the result of bad luck or bad policies in particular countries. Buiter and Srinivasan express this view when they declare that 'there is no "collective debt problem of the world". It is primarily a problem of a *few Latin American countries, that are relatively rich by the standards of developing countries.* . . .One is led to conclude that perhaps domestic mismanagement rather than external shocks explains the lion's share of the Latin debt accumulation' (Buiter and Srinivasan 1987, 413–14; emphasis in the original).

This view has been expanded upon in the press as well as in the academy. In various populist accounts of the debt crisis (as might be found in the *Economist* and *Fortune* magazines) the indebted countries are accused of waste and inefficiency on a grand scale; of a modern day Pharaonism. Mention is made of long roads to nowhere in Brazil, of unopened hotel blocks on a Mexican shoreline, of poorly constructed power plants and so on. Nepotism, too, and endemic corruption, are roundly condemned. Some problem debtors are portrayed as bottomless pits of avarice and incompetence, down which have flowed vast quantities of money from the developed world.

Other accounts are less stylized. Within a system-stability perspective it is common to focus attention on three particular and related instances of bad management: fiscal policy, capital flight and trade policy.

Consider, first, the question of fiscal policy. This is not only a concern of system-stability theorists; indeed, much of what will be said here will be accepted by system-instability theorists (albeit with a different gloss and with regard to different political horizons). The key point is that most countries in external debt difficulties have problems in balancing their books domestically. Countries which have avoided external debt problems have tended not to run fiscal deficits. The reasons for this state of affairs are a matter of dispute (see chapter 4). From a system-stability perspective the emergence of substantial fiscal deficits in Latin America and elsewhere is evidence of a local unwillingness to set strict monetary targets and then to abide by them. This reluctance in turn may be explained by reference to the nature of political regimes within rent-seeking societies. According to Anne Krueger and others, political regimes in countries which are not market- and rule-oriented are maintained in power either by repression or by purchasing political consent (Krueger 1974, 1985). Since such regimes are reluctant to increase local taxes, increases in public spending can only be financed by borrowing from home or abroad, and/or by resorting to the printing presses. In the 1970s the former option was clearly attractive to many MICs, and not least because of the very low real rates of interest charged. After 1981, external funds tended to dry up and governments were faced with a choice between belt-tightening measures or inflationary deficit financing. In too many cases the latter option was chosen. The sins of the 1970s – borrowing cheaply but unproductively – were continued into the early 1980s.

A simply money-balances approach to fiscal deficits links through to the phenomenon of capital flight. Again, it is a question of emphasis that matters here. Virtually all students of the debt crisis are agreed that capital flight has compounded the problems of the severely indebted countries. It is also widely accepted that capital flight is a direct consequence of large fiscal deficits and anti-inflationary policies of pegging the nominal exchange rate (after Sachs 1989b, 13). (Capital flight 'refers to the accumulation of foreign assets by the private sector of an economy, often at the same time that the public sector is incurring sharply rising external debts' (Sachs 1989b, 12). It tends to arise when governments increase transfer payments to the private economy in such a way that the exchange rate is weakened, thereby encouraging private holders of money to convert their holdings into foreign currencies to be deposited abroad. By its nature, capital flight is difficult to measure accurately.)

What is specific to a system-stability perspective is the suggestion that capital flight should be a matter of indignation for external creditors. This is not just because capital flight is symptomatic of lax fiscal policies; it is also because capital flight can be used as an argument in favour of debt repudiation, when it should signify the exact opposite. Peter Bauer puts this objection perfectly when he complains that: 'In 1987 the government of Peru had externally held reserves of about $1.5 billion, at a time when it refused to pay a few million dollars on servicing its sovereign debt' (Bauer 1991, 60; note that Bauer is further assuming that the government of Peru could or should requisition the overseas assets of its private citizens as well as of government parastatal organizations).

The phenomenon of capital flight in turn leads on to a discussion of trading regimes in developing countries. At this point a secondary division opens up in the ranks of system-stability theorists. According to Larry Sjaastad, the question of trade is very much subsidiary to the question of fiscal policy. He suggests that 'One of the key errors of some current thinking about the debt problem is that it is fundamentally a trade problem – one that will be greatly alleviated by recovery of the world economy' (Sjaastad 1983, 313; and compare Beenstock). In his judgement, if countries can 'generate the necessary *fiscal* surpluses they would probably have very little difficulty in producing the requisite trade surpluses as well. Fiscal surpluses are expenditure reducing and expenditure reductions are precisely what creates trade surpluses' (ibid. 315). Other scholars would put the emphasis more firmly on the question of trade and trade-related policies. Countries are enjoined to export their way out of debt.

Wherever the emphasis is put, system-stability theorists would agree upon the urgent need for problem debtor countries to move away from inward-looking development strategies. They would be

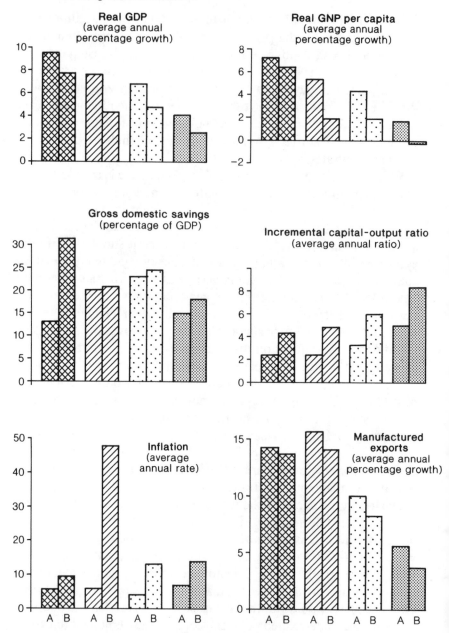

Figure 3.2   Macroeconomic performance of 41 developing
countries, grouped by trade orientation, 1963–1985
Source: World Bank (1987b, 84)

Strongly outward orientated

A 1963–73   and   B 1973–85
Hong Kong   Korea   Republic of Singapore

///// **Moderately outward orientated**

A 1963–73	B 1973–85
Brazil	Brazil
Cameroon	Chile
Columbia	Israel
Costa Rica	Malaysia
Cote d'Ivoire	Thailand
Guatemala	Tunisia
Indonesia	Turkey
Israel	Uruguay
Malaysia	
Thailand	

Moderately inward orientated

A 1963–73	B 1973–85
Bolivia	Cameroon
El Salvador	Colombia
Honduras	Costa Rica
Kenya	Cote d'Ivoire
Madagascar	El Salvador
Mexico	Guatemala
Nicaragua	Honduras
Nigeria	Indonesia
Philippines	Kenya
Senegal	Mexico
Tunisia	Nicaragua
Yugoslavia	Pakistan
	Philippines
	Senegal
	Sri Lanka
	Yugoslavia

Strongly inward orientated

A 1963–73	B 1973–85
Argentina	Argentina
Bangladesh	Bangladesh
Burundi	Bolivia
Chile	Burundi
Dominican Republic	Dominican Republic
Ethiopia	Ethiopia
Ghana	Ghana
India	India
Pakistan	Madagascar
Peru	Nigeria
Sri Lanka	Peru
Sudan	Sudan
Tanzania	Tanzania
Turkey	Zambia
Uruguay	
Zambia	

united in denouncing those policies which have discouraged the growth of tradable goods and services, and which have found expression in over-valued exchange rates. They would also insist that all policies in support of protectionism – in the developed world as in the developing world – are an offence against the maximizing principles of economic activity and theory. Restrictive trade policies, like unrestricted fiscal policies, are jointly responsible for specific local crises of debt and development. The World Bank sums it up in one of its excellent diagrams (see figure 3.2). The Bank argues that in inwardly oriented economies, the structure of incentives is so distorted that efficient resource allocation becomes difficult to pursue. On balance, trade liberalization policies are to be preferred because 'outward-oriented regimes provide self-correcting mechanisms to align the macroeconomic variables that affect growth' (World Bank 1987b, 90). Once again, it seems, we are returned to ideas of market efficiency, self-correction and unintentional equilibria.

5   Policing the Debt Crisis

Given what has been said so far, it will be apparent that there is no single SSP prescription for dealing with the so-called debt crisis. It is more accurate to speak of a range of proposals which are more or less consistent with this framework, and which call to mind the underlying logics of the counter-revolution in development theory and policy. What gives this range of proposals a certain unity is its willingness to discount other types of debt management proposals.

Probably the best way to make this point is to elaborate upon a form of reasoning set out by Professor Sjaastad in 1983. According to Sjaastad there are just three possibilities when it comes to determining who is to bear the losses (the blame) arising from bad debt:

(a)   The debtor countries can make the fiscal adjustments and reductions in general expenditure which are required to service their external debt.

(b)   The lending institutions can absorb the losses arising from their previous lending decisions.

(c)   Or the losses can be shifted, in whole or in part, to a third
       party, namely the taxpayers (or money holders) of the creditor
       countries. (Sjaastad 1983, 319).

System-stability theorists are as one in arguing for some combination of (a) and (b), on the basis of first denying the morality and
economic sensibility of option (c).

(1) System-stability theorists have three main objections to option
(c). Roland Vaubel is against this option because it will further
expand the role of the IMF in international economic affairs. (We
can assume that the IMF will be involved as an intermediary
between taxpayers on the one hand and direct creditors and debtors
on the other). In his judgement, 'there is no sound economic case
for IMF lending. . . .An international organisation that uses taxpayers' money to grant subsidised credits to governments and,
ultimately, banks because they have made mistakes in the past,
which they promise to correct in the future, is misconceived. . . .It
creates a moral hazard in international finance that would not
otherwise exist' (Vaubel 1983, 301).

   Others define moral hazard in less restrictive terms. Buiter and
Srinivasan suggest that option (c) is most likely to be put into effect
by proposals akin to the Baker and Bradley initiatives of the
mid-1980s (on the Bradley proposal see chapter 4). Such proposals
call for the further delivery of surplus funds to debtor nations from
the coffers of the industrialized world, including Japan. Buiter and
Srinivasan oppose this suggestion as follows. First, such a bail-out
would 'benefit those heavily indebted countries that are in serious
financial trouble (e.g. Brazil, Mexico and Argentina)' (Buiter and
Srinivasan 1987, 413); no such relief would be given to nations
which have serviced their debts (such as South Korea), nor will new
funds be made available to those countries too prudent or poor to
take out bank loans in the first place (like India and China). The net
result is that profligacy will be rewarded, which can only increase
the levels of moral hazard already present in the international
financial system.

   Second, such a bail-out would be to the advantage of those banks
which acted in an imprudent manner in the 1970s and early 1980s.
Buiter and Srinivasan are nothing if not consistent here. They are

not just supporters of option (a) on Sjaastad's menu; they are also supporters of option (b). They are opposed to that 'form of socialism for the rich much favoured by the banking community' (Buiter and Srinivasan 1987, 416). While the financial system should be protected by the central banks, and by the Federal Reserve Board in particular, it should not be 'beyond the combined wits of Paul Volcker and James A. Baker III to safeguard the integrity of the US payments and credit mechanism while letting the shareholders, creditors and managers of the commercial banks with the bad loans take the full measure of the losses caused by bad luck and/or by bad bank management. Continental Illinois, where the shareholders lost everything and the management was fired but the bank survived, is one example of what can be done' (Buiter and Srinivasan 1987, 416; see also Buiter et al. 1989). This is hardly conservatism!

Finally, and relatedly, there is the argument put forward by Peter Bauer. Bauer has little to say about the role of the money-centre banks when he writes about *The Third World Debt Crisis*. He is adamant, however, that 'many debtor governments have failed to meet their contractual obligations'; which is to say they have acted without honour and have 'defaulted in the proper, traditional sense of the word' (Bauer 1991, 56). Bauer argues that the problems of the indebted countries have arisen from 'policies that have wasted resources and damaged living standards and development' (ibid. 60). These policies, which include distortions to relative prices and efforts to restrict the inflow of private equity capital, have encouraged a degree of capital flight 'which is reflected in the very large external liquid funds of citizens of the major debtor countries' (ibid. 61). Bauer concludes that: 'What is known as the debt crisis does not arise from the inability of debtor governments to service their foreign debt without hardship to their peoples. It arises from the unwillingness of these governments to meet their obligations. In the present circumstances it is rational conduct for these governments to default' (ibid. 61–2).

Bauer's preferred policy response is in keeping with this conclusion, and perfectly exemplifies a libertarian strand within the system-stability perspective. Bauer suggests that Western governments and commercial lenders should make plain that there will be no further funds for countries in default on their debts. A ban on debt rescheduling will encourage a proper moral fibre in such countries. To date 'Western governments have condoned Third World default and have even encouraged it by insisting on forbearance by both official and commercial creditors in the face of default, including default on very soft loans' (Bauer 1991, 63). Bauer further suggests that the costs of the debt crisis must be borne by the debtor countries themselves and by the banks. Debtor countries took out loans on a voluntary basis and in the knowledge that interest rates can go up as well as down. If the West has a moral responsibility to these countries it is not to relieve them of any problems which they might now be facing. The West's duty is to encourage such countries to accept that individuals (and individual countries) are responsible for their actions and resultant fates. Non-market write-downs of debt, or debt rollovers, are offensive morally as well as being economically inflationary and counterproductive.

(2) The more positive proposals associated with a SSP are related to options (a) and (b) on Sjaastad's menu. Not surprisingly, these proposals tend to mirror some of the debt management practices which were pursued in the containment-and-adjustment years of the early to mid-1980s. Three main sets of proposals should be mentioned.

First, there is the question of adjustment. System-stability theorists follow two lines here. Some debt-cycle theorists tend to think of adjustment as a quasi-Darwinian process of adaptation and evolution. This view is also common among Austrian economists. Particular national economies, together with the global economy, are considered to be in a state of perpetual adjustment, with the required changes taking place more or less of their own volition. In this perspective, good market practices tend to drive out bad market practices.

A less sanguine viewpoint is taken by most profligacy theorists. Their focus is on those domestic economic rigidities which are an impediment to a desirable and natural process of market equilibration. Adjustment then becomes something to be struggled for, or to be actively sponsored. Insofar as domestic economic policies are inflationary and/or inward-looking – the product of rent-seeking regimes – external agencies are justified in pressing for processes of adjustment which will reduce these local scleroses. This is the case whether or not such agencies are, or should be, involved in involuntary lending practices to sweeten the pill of domestic reform. (This is obviously a moot point for system-stability theorists – cf. Vaubel). In empirical terms, adjustment means adjusting to the requirements of an open market economy. Budget deficits should be reduced either by increased taxation (which is not often favoured) or by public expenditure cuts. If these cuts have adverse effects on the lives of poorer people, this is an unfortunate but unavoidable circumstance. Just as economic actors must accept responsibility for their actions, so also 'There is no viable alternative to adjustment' (World Bank 1987b, 35).

Adjustment in the terms we have just described is considered to be a relatively short-term affair. The same is true of the containment strategy which is very much its twin. System-stability theorists are as one in resisting the suggestion that the debt crisis is a global crisis of solvency, as the Brandt Commissioners (among others) are wont to claim. The debt crisis is rather to be understood as a country-specific set of temporary liquidity crises which must be dealt with on a case-by-case basis. Insofar as involuntary lending is to be a part of the containment strategy, it should be pursued on a short-term basis and be so phrased that good performance is properly rewarded (see Claessans and Diwan 1990). System-stablity theorists are mindful of moral hazard in economic affairs and they are keen that partial forgiveness should only be given to those countries which have embarked upon a radical process of structural adjustment. Even then, the forgiveness should be negotiated on a voluntary basis. The same applies to bank write-downs of debt. System-stability theorists are sufficiently consistent to call upon the banks to play their part by recognizing bad debts. They do not believe that such a strategy constitutes a risk to the wider financial system.

Finally, there is the question of the longer-term. Having raised the issue of who is to bear the losses due on possible bad debts, Sjaastad ends his paper by asking: 'what reforms are required to reduce the probability of a similar situation arising in the future?' (Sjaastad 1983, 318). Besides the measures we have noted already, the prevailing wisdom in the system-stability camp is that problem debtors should reform their economies so that they 'get their prices right'. At the end of the day it is not indebtedness which is a problem for most developing countries; capital flows are actually to be welcomed. What matters is the ability of a given debtor to service its debts, and this depends on its ability to earn foreign exchange. The general advice, then, is for debtor nations to privatize parts of their economies, and to orient their economies to the global economy. At a minimum this will require the dismantling of protectionist barriers and the revaluation of local currencies. Once again, the logic appealed to is the logic of the competitive market. It is a discourse which was also advanced by the IMF and the World Bank for most of the 1980s, albeit in subtly different terms. We will return to this theme in chapter 6.

6 Conclusion and Critique

This chapter will not have been successful if it has given the impression that the system-stability perspective is a unitary perspective on the debt crisis, or that it is straightforwardly a creditor or IMF perspective. The views of a Bauer (or his mentor Hayek) are not always the same as the views of a Krueger or a Beenstock or a Vaubel. Still less is it the case that the work of Cline and Nunnenkamp is tied directly to a subjective preference paradigm, notwithstanding the fact that elements of their work are supportive of a system-stability perspective on international debt. The connections between ideas and theories and theorists are not quite this close. The subjective preference paradigm provides a series of resources – a series of 'ways of seeing' – which inform a counter-revolution in development studies, and which surface, variously, in a series of system-stability perspectives on the debt crisis.

Nevertheless, there is a degree of unity in the system-stability perspective, and this will become clearer as it is contrasted with two opposing perspectives (chapters 4 and 5). In a sense, it is the questions which tend not to be asked within a system-stability framework which best define it for us. Exactly what questions might be asked of it will become apparent in the next two chapters. At this point it will suffice to signal only the most general sets of objections which might be put to a system-stability theorist. They are four-fold.

First, what is the relationship between what might be called real time and abstract or model time in a subjective preference framework? Doesn't the system-stability perspective privilege an instantaneous account of market clearing, and a long-term vision of adjustment, over a more institutional and medium-term time scale which is more in tune with the concerns of real men and women? Less cryptically, isn't the SSP insensitive to the human costs of a process of adjustment which may take five to fifteen years?

Second, and relatedly, how are questions of space and regional linkages handled in a system-stability framework? Is it meaningful to theorize the debt crisis only on a case-by-case basis, and what are the implications of so doing? What are the unemployment costs in different places of place-specific processes of belt-tightening? Is there really no alternative?

Third, what is the relationship between banking stability and development stability in a system-stability framework? What, too, is the relationship between the private issuance of stateless monies and the possibility of system order (in an international relations sense)? Isn't the possibility of sound money management undermined by the offshore creation of international liquidity? Where does this leave the concept of general equilibrium?

Finally, what are we to make of the moral calculus advanced by the counter-revolution in development studies? Is it reasonable to take such a strong line on the indivisibility of the individual, and the individual's responsibility for his or her own fate? Are there no circumstances in which 'development' might be privileged over 'debt' in such a way that the needs and rights of distant strangers might be taken into account? Is it fair to penalize the poor in defaulting nations when many of them gained very little from the problem loans? Is it, indeed, reasonable to raise the issue of fairness, and if so how is it to be raised? Where does this leave the question of intentionality?

Answers to some of these questions, together with different readings of the debt crisis, will be found in the next two chapters.

Notes

1   If Nozick (1974) offers the most consistent exposition of an entitle-
ments approach to questions of economic justice and distribution, a
thoughtful recent book by Israel Kirzner (1989) puts forward an ethics
of finders-keepers which is more in accordance with the traditions of
Austrian economics. Kirzner is sceptical of social justice arguments
which assume that a given pie exists to be divided up according to
some agreed set of rules or social contract (cf. Rawls 1972, and see
chapter 4). He proposes that that the slicing up of an economic pie
cannot meaningfully be decided upon or discussed except in relation to
the process of its production. This process of production, Kirzner
argues, depends fundamentally upon the entrepreneurial act of discovery amidst the circumstances of radical uncertainty which characterize
free markets. Kirzner maintains that 'What is needed, therefore, in a
theory of economic justice, is an approach that recognizes that at each
stage in capitalist economic activity what is won has been, in some
degree, found. Whether we deal with the justice of resource ownership
or the source of just title in produced output, we are dealing with the
ethics of assigning that which has been discovered. There is every
reason to believe that, in the judgement of many ethical observers, it
matters a great deal that we are concerned with the assignment of
discovered, created, products, rather than with economic goods that
are seen, essentially, as having been known to be here from the
beginning of time' (Kirzner 1989, 16).
2   The comparison that Buiter and Srinivasan make between Argentina
and South Korea is not entirely innocent: Brazil, Chile, Mexico and
Venezuela all had a far greater percentage of their external liabilities at
floating rates in 1984 than did South Korea.

FOUR

The Debt Crisis: a System-correction Perspective

1 Introduction

Most interpretations of the developing countries' debt crisis do not derive from a system-stability perspective or from a system-instability perspective. They have their roots in a perspective that is more difficult to pin down and to place boundaries around and within: the system-correction (SC) perspective.

The SC perspective includes within it scholars and practitioners who are sympathetic to the view that developing economies should become more open, while remaining sensitive to the difficulties this must present in a global economy which is far from open and symmetrical. SC theorists also refuse to conflate a prospectus for openess in international affairs with *laissez faire*, or with a singular faith in *the* market as an abstract entity. Markets, in this perspective, are usually desirable but often imperfect institutions. They are constituted according to particular sets of relationships which are more or less asymmetrical and more or less transparent. As such, real markets are always prone to failure and to macroeconomic disequilibria, at least in the short to medium term. This is also true in the case of international finance. The SC perspective emphasizes the debilitating effect of developing country debt problems on international economic and political affairs, as measured by rates of unemployment and inflation, by social unrest and by a series of development indicators. This perspective accepts that the debt crisis is a common crisis which demands a common set of solutions. It may also signify a local or a global crisis of solvency.

Thus described, the system-correction perspective is consistent with a broad range of economic and political philosophies and with a variety of interpretations of the developing countries' debt crisis. More so than ever, it is important in this chapter not to be seduced into the belief that the SC perspective is a unitary perspective. What lends this perspective a certain distinctiveness is its focus upon macroeconomic relationships and disequilibria and its unwillingness to place blame upon only one or two parties to a state of prospective economic breakdown. Pragmatism is here the watchword.

The rest of this chapter seeks to build upon this rather thin definition. Section 2 offers a more considered analysis of some of the economic and philosophical strategies and claims that are consistent with a system-correction approach to debt and development. Important among these strategies are Keynesian and post-Keynesian economics, and pragmatic accounts of freedom, justice and responsibility. These claims are sufficiently diverse that they do not always make the best of companions. Debate and dissent within the Keynesian-pragmatist paradigm is quite likely and reasonable (and quite likely to be reasonable).

Section 3 considers the relationship between Keynesian-pragmatism and what is sometimes called Keynesian development studies, or duoeconomics. After noting certain problems with this nomenclature, this section considers how an early faith in capital-based growth models has given way to a more eclectic set of formulations on the mixed economy in developing countries. In part, this shift is in recognition of the challenge to duoeconomics mounted by the counter-revolution in development studies; in part, it is a recognition of the changing nature of the world economy post-1970. An important and continuing tradition of dissent within Keynesian development economics is the tradition of structuralist economics, and this is briefly outlined.

Section 4 considers how a system-correction perspective makes sense of the origins and significance of the debt crisis. Rather than re-hashing the standard narrative account of the debt crisis which is at its core, our focus is on those interpretations of the debt crisis which diverge markedly from rival sets of interpretations. This means that a particular emphasis is placed on the prospects for world economic growth and its likely impact on indebted regions;

on the morality of structural adjustment policies; and on the political context for economic management.

Section 5 outlines a range of proposals for debt crisis management which have emerged from within a system-correction perspective. These proposals can be more or less radical (as in the case of some structuralist suggestions), and more or less restrictive (as in the case of the proposals put forward by Rohatyn and Kenen). Section 6 offer a brief critique of some aspects of the SC perspective.

2   Keynesianism–pragmatism

Most system-correction accounts of the debt crisis are indebted to the traditions of Keynesianism and pragmatism. These links are not always very direct and this contention depends on the suggestion that Keynesianism is more than just a recipe for state intervention in the economy. Keynesianism is also about forms of economic reasoning which treat uncertainty and aggregation in a manner which is not always evident in the subjective preference paradigm. Keynesianism also contains within it voices of dissent. In this respect it is similar to the broad-ranging traditions of pragmatism to which it can be allied.

Keynesianism

One way to make sense of Keynesianism is to place it in historical context. Keynesian economics has been described as that version of economics which aims 'to save capitalist society from the dangers posed by rising unemployment and falling wealth' (Wolff and Resnick 1987, 101). This judgement is a little unflattering, but one sees the sense of what is being said. Keynes's great work on *The General Theory of Employment, Interest and Money* was first published in 1936 and its clear point of reference is the UK economy amidst the depression of the 1930s (Keynes 1936). Like Marx and Schumpeter, Keynes was concerned that market capitalism could be associated with a secular stagnation which might induce political instability. He also joined with these authors in looking beyond the seemingly apolitical traditions of neo-classical economics. Where

Keynes parted company with Marx and Schumpeter was in his attempt to marry the economic insights of Marshall to the traditions of political economy associated with Ricardo and Malthus. Keynes was minded to argue in favour of certain 'corrections' being made to a market-based system of capitalism. Although Keynesianism would later be denounced as a charter for economic serfdom (Hayek 1944), Keynes did not doubt that markets had to be defended and that open markets were an important basis of an open society. More so than some of his followers, Keynes was a liberal in the American sense of that word.

In this book it is not necessary to review Keynesian economics in great detail. It will suffice to present Keynesianism as an economic hybrid which is sufficiently distinctive from subjective preference theories to be taken seriously in its own right. We also need to recognize that Keynes was not the sole architect of what was to become 'Keynesianism'. Keynes's work at Cambridge was in part anticipated by the work of his colleague-to-be, Michal Kalecki, albeit in more radical terms (Osiatyński 1990a,b). What is called Keynesianism has also been refined by three generations of Keynesian, neo-Keynesian and post-Keynesian economists. In the 1990s it will be more difficult than ever to identify where Keynesianism ends and where other traditions of economics begin.

This does not mean that Keynesianism is so elusive as to defy description. If we return to the work of Keynes himself, we find there four sets of propositions which anticipate the later discourses of Keynesianism.

A first set of propositions concerns the characteristics and settings which define the economic agents active in the Keynesian economic landscape. These characteristics and settings are subtly different to those found within a subjective preference paradigm. In place of the free-standing actor who seeks to maximize his or her utility, we are introduced to much more problematic characters. Keynes talks of mass psychologies and of animal spirits. In his world, actors are always socialized actors; men and women who are rational only in regard to certain long-standing conventions or cultural norms. One such norm or expectation is that workers in mature capitalist economies will not willingly accept lower money wage rates for the tasks which they perform. It might be that unemployed workers could price themselves back into work if they were willing to accept a rate

of pay 60 per cent below that which is generally considered acceptable. In most circumstances, however, the mass psychology of the
working class is such that this option is not considered. One consequence is that money wages tend to be sticky in a downwards direction. This will be especially the case where workers are lent the
support of collective institutions such as trades unions.

This leads to a second set of observations. Having noted that
workers are not inclined to act in a manner which is consistent with
the predictions of orthodox or neo-classical economic theory,
Keynes suggests that an economy can come to rest at a state of less
than full employment. In the short term some workers are unwil-
lingly unemployed and economic activity in general is depressed.
This in turn affects the climate of business confidence in an
economy. Investments will tend to fall as individuals choose to hold
their assets in more liquid forms. The result is an economy which is
depressed even further.

Keynes puts this more precisely, of course, and his account of
labour market failures is part of a much wider critique of the
unambiguous equilibrium solutions proposed in most neo-classical
models. In the case of savings-versus-consumption behaviour,
Keynes insisted that the amount saved could not be assumed to be
independent of the amount invested. In other words, savings and
investment decisions do not simply depend upon the rental rate of
capital, and nor does increased investment necessarily call forth
increased savings by 'changing the price of future relative to current
consumption' (Wolff and Resnick 1987, 114). In the case of money
market transactions, Keynes insisted that individuals have a
psychological propensity to hold money for liquidity and speculative purposes, as well as for making transactions. The total demand
for money in an economy then becomes a function 'not only of real
income [as in the quantity theory of money]. . .but of the rental
rate as well, because of speculative or liquidity needs' (Wolff and
Resnick 1987, 115). Putting these propositions together, Keynes
concluded that real output and employment are determined by
spending; for the neo-classical economist, by contrast, it is employment which determines the level of real output in an economy.

Keynes's third set of propositions derives from this last point.
Insofar as there is 'no reason to expect investment demand to
increase when business prospects are. . .poor' (Wolff and Resnick

1987, 116) it follows that aggregate spending must be increased by the state. The state is now called upon to fine-tune the economy by means of fiscal and monetary policies, in addition to providing the legal and property regimes necessary for the functioning of capitalist societies. This is what is commonly understood by Keynesianism (Hall 1989). This set of propositions is a target for those critics who like to equate Keynesianism with *dirigisme*, or with an unhelpful meddling in the more elemental affairs of the market.

Finally, there is the matter of Keynesianism as economic philosophy. It will be clear that Keynesianism is not committed to a tradition of methodological individualism in the same way that neo-classical economics is. Keynesians are prepared to place economic actors in some sort of social and political context, and to deal straightforwardly with imperfect collective institutions such as trade unions and monopolistic firms. The Keynesian world of markets is also a world of uncertainty (but not in a Hayekian sense). In the Keynesian world-view, markets fail notwithstanding the fact that individuals might be acting rationally. Keynes suggested that privately rational actions do not always combine to produce an outcome which is optimal from the point of view of society. The so-called 'isolation paradox' will be especially in evidence when entrepreneurs act for short-term reasons and in order to maximize speculative gains. It also follows from the absence of perfect information within an economy. Wolff and Resnick point out that 'one effect of this "natural" phenomenon [uncertainty] is the inevitable possibility of a disruption between the more or less stable savings in a society and the levels of investment rendered volatile by uncertainty. In this sense, declines in investment spending *are not anyone's fault*; their cause is ultimately reduced to our imperfect human nature' (Wolff and Resnick 1987, 121; emphasis added).

This last point is very well put, and it leads on to a consideration of the ways in which more abstract questions of space and time are handled in Keynesian and post-Keynesian economics. In the work of subjective preference theorists time is often theorized as a point or an instant, or in terms of some more slowly working processes of economic adjustment. For Keynes, meanwhile, 'in the long run we are all dead'; by which he means that real men and women are more concerned about what happens in the short and medium term. By the same token, space in most Keynesian models is defined extensively, and with due regard for national and regional boundaries and systems of cultural expectations. The more abstract and/or point-based conceptions of space to be found in neo-classical economics are not evident here. Keynesian economists recognize the spatial interdependencies present in contemporary economic systems, and they tend to be sensitive to the effects in one set of locations of processes of economic adjustment pursued in another set of locations. The same cannot always be said of the work of subjective preference theorists. Keynesian economists are also more likely than their neo-classical counterparts to respect the longevity of those political and cultural systems within which economic reforms must be pursued. Although most modern Keynesians are disposed to welcome reforms which further the production of open and competitive markets, they are often unwilling to allow this preference to override a sensitivity to the durability of social practices.

Wolff and Resnick conclude that these contrasting outlooks mark Keynesian economics as a discourse which is uncertain of its own propositions. In their view Keynesian economics is neither fish nor fowl, but rather an uneasy alliance of abstract market principles and political common sense. This may be so, but a more charitable reading would suggest that Keynesianism is here displaying its abiding pragmatism. Its protagonists see markets which are in need of correction, and propose that corrections be made notwithstanding the fact that the corrections might later have to be corrected. In a world which is always uncertain and imperfect, a willingness to dispense with abstract principles can be both refreshing and rewarding.

Pragmatism

The claims of pragmatism offer a second strand to the Keynesianism–pragmatism perspective we are describing. From what has been said thus far, it might be apparent that Keynesian economics defines a continuum of economic models and proposals which runs from a neo-classical–Keynesian synthesis (the new classical economics) to versions of structuralist and institutionalist economics which are much less inclined to place their trust in general equilibrium models. The political and moral philosophies which might be associated with Keynesianism are similarly broad. Whether these philosophies can reasonably be lumped together under the heading of pragmatism is a moot point. It will, after all, mean talking about conservative paternalism and a Rawlsian theory of justice in the same breath. Nevertheless, this is the tack which will be taken here.

We can start with Keynes himself. We have noted that Keynes was a staunch supporter of a managed capitalism, and certainly he was a tireless proponent of the rights of the individual. At the same time, Keynes was mindful of the power to act selfishly which is born of the animal spirits to which he often referred (Harrod 1972). At a national level, Keynes believed that selfish self-deception was typical of the attitudes of the English and French establishments at the time of the Versailles Peace Conference in 1919. Forcing the Germans to make heavy reparations to the English and French would not be in the medium-term interests of European economic stability (Keynes 1971). It is also significant that Keynes expressed the hope that the 'money motive' would become less pressing in a system of managed capitalism.

Other scholars have moved beyond Keynes in different directions. To the right of Keynes are scholars and policy-makers who believe that the distribution of rewards in society is justified in economic terms, but who suggest that the politics of envy can be destabilizing politically. The pragmatic solution to this dilemma is one of limited state intervention. Governments would be charged with the task of providing minimal economic citizenship rights for the least well off in a country. Others might arrive at this view (which overlaps with the views of some subjective preference theorists) by means of a pragmatism born of conservative paternalism. In this case there is a general presumption that the better off in society should help those less well off than themselves. Once again, this is a question of attending to the needs of strangers, as opposed to recognizing any rights that the poor might have to a larger slice of the economic pie.

Still other scholars have taken up positions to the left of Keynes, while subscribing largely to Keynesian tenets about the economy. The followers of Michal Kalecki, for example, would be less willing than most Keynesians to endorse the deep-seated virtues of market capitalism. Kalecki's work bears the imprint of a dialogue with

Marxism, and his models of crisis formation and displacement under capitalism are correspondingly more sensitive to questions of class than are the models developed by Keynes. Kalecki is also less prepared than Keynes to trust the economy to a neutral and technocratic state. Kalecki suggests that a redistribution of social assets will be a pre-condition for effective state intervention in the workings of a market economy. In less radical contexts the government will be held hostage to the private demands of the proprietary elites that assume power within a state.

Although Kalecki's work is not attached to any one political philosophy, his work has sometimes been cited by those who are keen to argue for a market or feasible socialism (see Nove 1983). It may not do too much violence to his views, however, to suggest that Kalecki's version of Keynesianism is complemented by the theory of justice developed by John Rawls (1972). This might sound strange because Rawls is a social democrat and his theory of justice is at least as supportive of the social market economy as it might be of market socialism. In any case, the Rawlsian theorem cannot be linked to determinate political outcomes in any precise way. Nevertheless, the Rawlsian theory of justice is radical in at least some of its propositions. Although Rawls privileges a principle of liberty over a principle of difference, the latter principle is in stark contrast to the proposals associated with the radical right. The Rawlsian difference principle suggests that 'social and economic inequalities are to be arranged so that they are both (a) to the greatest benefit of the least advantaged, and (b) attached to positions and offices which are open to all under conditions of fair equality of opportunity' (Miller 1991, 423).

In more straightforward terms, Rawls's social contract theory of justice is based on the contention that 'there but for the grace of God go I'. Rawls makes his case by means of an appeal to 'the original position', or what he calls 'the veil of ignorance'. Imagine, says Rawls, that the principles of a just society are to be drawn up by agents who are 'deprived of knowledge about their talents and abilities and about the place they occupy [and will occupy] in society' (Miller 1991, 422). Imagine, in other words, that the rich and the powerful are not already the rich and the powerful, and that the chance of becoming one of the rich and the powerful is far less than the chance of becoming one of the poor and dispossessed. In such a situation – where on present odds one's chances of being born into a middle-class family in America are much less than the chance of being born into destitution in the developing world – the pragmatic actor would want to (re)-arrange society so that at least minimal standards of freedom and livelihood are guaranteed for all, *and as of right*.

In this manner, Rawls provides a theory of justice from first principles. It may not be a theory of justice which is easily acted upon, and it may be that the Rawlsian difference principle is not meant to extend to an international stage. That said, the Rawlsian model is a reasoned command to think that the needs and rights of strangers could easily be the needs and rights of ourselves. Rawls's account of what it means to be an individual is an important one. The Rawlsian model also offers an alternative to the 'just deserts' models so liked by subjective preference theorists. The Rawlsian world, like the Keynesian world, is a world of uncertainty and self-doubt. Rawlsian individuals are encouraged not just to count their blessings, but to recognize that any blessings they might have – money, wealth, talent – are largely the products of fate (including the accident of birth). To the extent that governments can specify a social welfare function, it is then legitimate for the state to redistribute wealth from one group of citizens to another. We will return to the logic of this conclusion later in the chapter.

3 Keynesian Development Studies

If it is difficult to link Keynesianism directly to a political philosophy, it is just as difficult to write of Keynesian development studies. Indeed, it is remarkable that the terminology itself tends only to be accepted by those hostile to its supposed intentions; that is, by the counter-revolutionaries and the Marxists. Most development economists who would seem to be pursuing a Keynesian line of argument resist this suggestion. They point out that Keynes himself was not much interested in the problems of developing countries, and that leading 'Keynesian' development economists, such as W. A. Lewis and Raul Prebisch, have denied that their first debt is to Keynes. Nevertheless, it is possible to identify a broad-based Keynesian development economics – especially if we are prepared to look for the indirect effects of an intellectual tradition upon particular economic and political propositions.

Diane Hunt comes to this conclusion in her survey of different economic theories of development (Hunt 1989, 27). She follows Hans Singer in suggesting that a Keynesian influence on development studies has been significant in five key respects: (a) in terms of Keynes's willingness to provide more than one economic model, to take account of differences in economic systems and levels of unemployment; (b) in terms of Keynes's attention to policy issues in the broad area of macroeconomics; (c) in terms of Keynes's impetus to the development of national income accounting and the collection of macroeconomic statistics; (d) in terms of Keynes's guarded recognition that protectionism can be a legitimate means to securing full employment and output in a national economy; and (e) in terms of Keynes's proposals for setting up an international bank with the power to issue international forms of liquidity (*Bancor*, later to be re-born as the Special Drawing Right).

Within these parameters we can talk of a Keynesian development economics which comprises two long-standing traditions. There are, first, the early Keynesian growth models, or what Hunt calls the expanding capitalist nucleus models. Second, there are the structuralist models associated with economists working for the United Nations' Economic Commission for Latin America (ECLA). These latter models marry Keynesian economics with more radical economic traditions, while the capital-growth models are more willing to advertise their underlying attachments to free markets and to entrepreneurship. Finally, there are some more recent developments within mainstream 'development economics' which are worth noting.

Expanding Capitalist Nucleus Models

Central to the expanding capitalist nucleus tradition are various books and papers published between 1945 and 1965 by Rosenstein-Radan, Rostow, Nurske, Lewis, Liebenstein and several others (Rosenstein-Radan 1943; Nurske 1953; Lewis 1955; Liebenstein 1957; Rostow 1960). These early models were later refined and given subtle new inflections by scholars and practitioners including Chenery, Robinson and Syrquin; Fei, Ranis and Kuo; and

Johnston and Kilby (Chenery et al. 1986; Chenery and Syrquin 1975; Fei et al. 1980; Johnston and Kilby 1975). What the models tend to have in common is a willingness to equate economic development with a process of structural transformation. Structural transformation is defined as a process whereby a majority of the agricultural workforce of a given country are transferred to more productive jobs in the non-agricultural sector of the economy. The models also suggest that the process of structural transformation can be speeded up to the extent that 'latecomer' or developing societies can adapt innovations and customs which were supportive of a prior process of development in the advanced or 'pioneer' countries.

What this process of structural transformation most depends upon is the rate of savings within a country. Savings are widely taken to be the key to development (by means of local processes of industrial capital formation), and savings can be enhanced by government policies and by injections of foreign aid and foreign direct investment. Savings will also be maximized to the extent that governments create the conditions for the formation of an entrepreneurial class. This is where most counter-revolutionaries are off the mark when they seek to criticize a corrupting development economics. If one returns to the work of Rostow, one finds there a clear assumption that the promotion of capitalism is a pre-condition for a country seeking to take off into self-sustaining development. Rostow's classic work, after all, was subtitled *A Non-Communist Manifesto* (Rostow 1960). Although not all modernization theorists take the same line that Rostow espoused, the general premise of mainstream development economics in the 1950s and 1960s was that inequalities of income within a developing country would first have to widen to allow the entrepreneurship for economic growth which would make possible a later redistribution-with-growth.

Finally, the expanding capitalist nucleus model is identifiable by its optimism with respect to development. The imagery of the Rostow model is especially powerful in this regard. For Rostow, development is about the technological frontier and the conquest of old spaces and old habits. His model is full of orbital imagery, with satellite countries taking off into an era of self-sustaining development. But Rostow is not alone in this view. The Development Decade proclaimed by the United Nations in 1960 called to mind a timetable of development which might be numbered in years rather than decades. As long as governments maximized the resources available to them, and directed those resources to key industrial sectors, development would come swiftly to the countries concerned.

Structuralist Models

A second tradition within Keynesian development studies is less optimistic than the expanding capitalist nucleus paradigm. Scholars and practitioners who have expounded a structuralist line on development have argued in favour of industrialization as a mainspring to development (like modernization theorists), but have suggested that industrialization is unlikely to materialize within the structures of domestic and international economic relations which presently maintain. Raul Prebisch and Celso Furtado are just two ECLA economists who have argued strongly against a simple equation of economic growth with economic development (Prebisch 1951; Furtado 1963). Their focus has been on the building up of developing economies as broad-based economies, in which all branches of production adopt advanced production technologies. Their focus, in other words, is on a balanced growth which will require the coordinating actions of an over-reaching economic agency, the state.

The structuralist perspective further argues that the underdevelopment of most developing countries is a consequence of their uneven and asymmetrical integration into the world economic system. The present and ex-colonies had their economies distorted to serve the needs of their economic and political masters. Primary commodity production has been over-emphasized and domestic manufacturing industry was positively discouraged. The result was a periphery dominated by external interests and by backward-looking domestic landed and trading interests; a group of economies unable to develop as long as they were dependent upon the export of goods which tended to suffer from a secular decline in the terms of trade. Put bluntly, the ECLA diagnosis was a diagnosis within the Keynesian tradition, but at one end of it. ECLA called for a new international division of labour and urged Latin American countries to help construct this new order by first setting up consumer-goods based import-substituting industries. In India, a similar recipe for protectionism and infant-industry support was predicated upon the building up of capital-goods based industries. In both cases, domestic and international terms of trade had to be 'distorted' to rid a country of the economic distortions born of colonialism and neo-colonialism. *Dirigisme* was the practicable option for development in an unequal world economy.

Dilemmas of Development

A footnote to this section should be briefly entered. On the basis of some critiques of an impoverishing 'development economics', one would think that Keynesian development studies has not moved on since the early 1960s. When *dirigisme* for development is attacked by the counter-revolutionaries, it is almost always with the work of Lewis, Nurske and Prebisch firmly in mind. This method of critique has some validity, insofar as mainstream development studies continues to attach itself to such totems as foreign aid, industrialization and partial state intervention in the economy. It is a misleading form of critique, however, insofar as it ignores more recent advances within 'Keynesian' development studies. These advances have been made in response to earlier critiques, and in recognition of certain changes in the nature of the world economy. Mainstream development economics is now more sensitive than it once was to questions of microeconomic efficiency in the developing world, and to issues relating to open trading regimes and human capital formation (Toye 1983). Within the more radical reaches of Keynesian development studies there is also a greater concern for the basic needs of individuals, and for development projects which put the poor first (often making use of indigenous resources and knowledge systems), rather than for those which triumph industrialization as a panacea for development.

In short, it is not especially helpful to write as though mainstream development studies, uniquely, is attached to propositions which may or may not have been sensible thirty or forty years ago, but which are out of date in the 1980s and 1990s. Nevertheless, it is the case that mainstream development economics continues to exhibit an abiding pragmatism. Most of its proponents are interested in exploring the ways in which states and markets, external and internal agencies, can be combined in particular places to particular effects. Instead of making plain a more fundamental faith in *the market*, 'Keynesian' development economists prefer to work through those imperfections and trade-offs which define the real-world dilemmas of development.

4 Debt and Development

If a willingness to think and act in pragmatic terms is characteristic of Keynesian development studies, so also is it a characteristic of a system-correction perspective on the developing countries' debt crisis. Most system-correction theorists advance an account of the debt crisis which is consistent with the standard narrative account set out in chapter 2. In other words, the debt crisis is not assumed to be the fault of any one set of actions, mistakes or circumstances. Nor is it assumed that a problematic future could easily have been foreseen or guarded against; wisdom and the benefit of hindsight are not to be confused. Finally, the debt crisis is placed into a political framework which highlights the need for political consent for economic policies.

Rather than repeating the main claims of the standard narrative account of the debt crisis, the rest of this section is organized as follows. First, attention is given to some of the more important analyses of the debt crisis made from within a system-correction perspective. The work of Sachs, of Díaz-Alejandro, and of Griffith-Jones and Sunkel is outlined to give a sense of the range of views consistent with this perspective. Second, a critique is developed of the containment and adjustment strategies for debt crisis management which were pursued throughout most of the 1980s. This critique is developed to highlight certain common points of focus within a system-correction perspective.

System-correction Models of the Debt Crisis

System-correction theorists vary in the emphases which they choose to place on different parts of the standard narrative account of the debt crisis. Some SC theorists are working close to the borderline with a system-stability account (see the work of Jeffrey Sachs);

others, like Díaz-Alejandro and Griffith-Jones and Sunkel, are exploring mainstream and structuralist narratives within the SC perspective.

Jeffrey Sachs

Jeffrey Sachs is a Professor of Economics at Harvard University and a research associate of the National Bureau of Economic Research (NBER). In his latter capacity he directed an important research project for the NBER on 'Developing country debt'. The findings which emerged from this study have been published in a series of books and papers, including an excellent collection of essays on *Developing Country Debt and the World Economy* (Sachs 1989a). Sachs has also published widely on the debt crisis in his own right (Sachs 1986, 1987, 1988).

Sachs's perspective on the debt crisis has shifted subtly over the years and in the light of his experiences as an economic advisor to problem debtor countries, including Poland, Bolivia and Russia. Sachs is sceptical of 'mainstream creditor interpretations' of the debt crisis on the grounds that they confuse free-market policies with open economic policies (see later). They also conflate policy mistakes evident with the benefit of hindsight with policy mistakes which could have been identified at the time of enactment. He also mistrusts the suggestion that granting debt relief to some debtors 'would hurt the debtors more than it would help them' (Sachs 1989b, 5). By the same token, Sachs is not impressed by the debtors' perspective on the debt crisis. This perspective puts most of the blame for the debt crisis on the fiscal and monetary policies pursued by the United States in the late 1970s and early 1980s and on various exogenous shocks to the economic systems of the developing world. Sachs notes that this perspective is insensitive to the different ways in which developing country debtors coped with external shocks and US inspired policies. Domestic policies are not given sufficient attention.

More positively, Sachs maintains that 'The debt crisis arose from a combination of policy actions in the debtor countries, macroeconomic shocks in the world economy, and a remarkable spurt of unrestrained bank lending during 1979–1981' (Sachs 1989b, 6). When Sachs refers to poor policy actions he has in mind chronically large budget deficits, over-valued exchange rates and a trade regime which discriminates against exports (and agricultural exports in particular). All this is unremarkable, as are Sachs's proposals for dealing with the debt crisis: faster economic growth in the world economy; structural adjustment in the developing countries following a financial time-out for problem debtor countries; and the further development of financial instruments which can enhance local processes of debt forgiveness and debt writedowns. What makes Sachs's work distinctive in his attention to detail (in the case of US bank lending and US banking regulations: Sachs and Huizinga 1987), and his persistent emphasis on the debilitating role played in the debt crisis by capital flight. Sachs is sceptical of the claim that bank lending to Latin America dried up suddenly in 1982 as a result of the Falklands-Malvinas conflict and a subsequent bankers' panic. Sachs suggests that 'The cutoff arose as much from a remarkable hemorrhaging of dollars from these economies, in the form of capital flight, after 1980. Foreign official borrowing by the Latin American economies supported perhaps $50–60 billion of capital flight in 1981–1982 alone' (Sachs 1984, 395).

This is an important point and one that puts Sachs closer to the views of some system-stability theorists than is common among system-correction theorists. Sachs illustrates his case with reference to Venezuela. He points out that 'From 1974 to 1982, Venezuela ran a cumulative current account surplus of $5 billion. It enjoyed two huge terms of trade gains in the decade during the oil shocks of 1973–74 and 1979–80. By 1981, it had accumulated foreign reserves of $19 billion. And yet by 1983, real GNP was falling by 4.7 per cent and the government is now [1984] renegotiating $22 billion of external public debt' (Sachs 1984, 395). Sachs's point, of course, is not that problem debtors are wholly to blame for the situation in which they find themselves. He simply argues that central bank support for over-valued exchange rates in some Latin American countries allowed capital flight to grow to alarming proportions in the early to mid-1980s. By 1981–2 some Latin American countries were borrowing publicly to support a corresponding build-up of net private assets abroad.

Although Sachs does not develop the political context in which capital flight manifests itself as clearly as does Díaz-Alejandro, he does note that the situation in South-east Asia is very different to the situation in Latin America. Sachs also points out that the comparative absence of capital flight from indebted South-east Asia has very little to do with the free-market trade regimes supposedly pursued within this region. It is testimony rather to the strong actions of some South-east Asian governments (including the government of South Korea) in blocking free convertibility in US dollars by means of capital controls. The lesson from South-east Asia is thus not that liberal economies perform better than *dirigiste* economies, at least not in any simple way. The lesson is that economic performance depends crucially upon the nature of the support which governments are prepared to offer to export-oriented industries, and in regard to local regimes of capital inflows and outflows (Wade, 1984).

Carlos Díaz-Alejandro

A second version of the system-correction perspective was developed by the late Carlos Díaz-Alejandro, particularly in his classic paper 'Latin American debt: I don't think we are in Kansas anymore' (Díaz-Alejandro 1984; see also Findlay 1988). Díaz-Alejandro is more inclined than Sachs to blame the debt crisis upon unforeseen and exogenous shocks to the economies of developing countries. A number of the points that he makes are echoed in the work of Ajit Singh (Singh 1988) and Albert Fishlow (Fishlow 1985; Cardoso and Fishlow 1989).

Díaz-Alejandro begins his 'Kansas' paper with a blunt declaration. 'Blaming victims', he says, 'is an appealing evasion of responsibility, especially when the victims are far from virtuous. But when sins are as heterogeneous as those of the Latin American regimes of 1980 one wonders how well the exemplary mass punishment fits the alleged individual crime' (Díaz-Alejandro 1984, 335).

Díaz-Alejandro argues his case as follows. He first presents a series of debt and development indicators for six Latin American countries which pursued very different domestic and external economic policies in the 1970s and early 1980s; Argentina, Brazil, Chile, Colombia, Mexico and Venezuela. Díaz-Alejandro notes that 'all six [major debtor countries] had serious economic difficulties during 1982–83 and faced a weak recovery during 1984' (Díaz-

Alejandro 1984, 336). This was as true of a militantly *laissez faire* country like Chile as it was of 'decidedly interventionist' Brazil and what Paul Krugman calls 'wacko Argentina' (Krugman 1984, 392).

Second, although some opposition parties in Brazil and elsewhere complained about the 'pharaonic projects' which were financed by external debt in the 1970s, there were 'few documented criticisms casting serious doubt on the *ex ante* social profitability of specific projects; even fewer criticisms documented at a disaggregated level a decline in investment efficiency between 1966–73 and 1973–1980' (Díaz-Alejandro 1984, 340). Díaz-Alejandro rightly emphasizes this point. He notes that most creditor interpretations of the debt crisis are keen to conclude that the problem debtor countries made horrendous policy mistakes of omission and commission in the 1970s and early 1980s. At the time, however, representatives of these same creditors failed to recognize just those 'mistakes' which they would point to some years later. In the 1970s, indeed, investment in dams, hydro-electric projects, railways and so on was recommended as a sensible use of funds.

Third, exchange rate policies varied throughout the six countries and failed to predict the likely consequences of specific trading patterns. Fourth, by 1980, 'debt indicators presented a mixed but not necessarily an alarming picture' (Díaz-Alejandro 1984, 342). Díaz-Alejandro points out that information on indebtedness was imperfect before 1982, that capital flight from Latin America was not massive in the early 1980s (here he parts company with Sachs) and that banks could reasonably have been expected to continue lending to the indebted developing countries. Some support for this view comes from the World Bank. In its first *World Development Report* of 1978, the World Bank noted that the debt ratios it was predicting up to 1983 were 'not unacceptably high. . .and should pose no general problem of debt management provided exports can grow at the projected rates' (World Bank 1978, 31).

Fifth, while some Latin American countries were in urgent need of reform policies in 1980–1, 'nothing in the situation called for traumatic depressions' (ibid. 348). Díaz-Alejandro suggests that the commercial banks acted precipitately in pulling the plugs on net new lending to Latin America in 1982–3. Sixth, it was this that sparked off the debt crisis. The collapse in net inflows in 1982–3, combined with a decline in the absolute dollar value of exports prior to the decline in loans, meant that a debt crisis was inevitable in Latin America in general, and regardless of domestic economic policy mixes. The comparison with South Korea is thus unwarranted. South Korea is a one-off; the more meaningful basis for comparison lies within Latin America itself.

Díaz-Alejandro concludes with some remarks on debt crisis management. Perhaps rather surprisingly, given the tenor of his remarks, Díaz-Alejandro calls for a range of proposals which are similar to those endorsed by Jeffrey Sachs – a rather eclectic mix of 'better exchange-rate management in the debtor countries, moderate IMF policies, faster OECD growth and modest systemic reform', as Sachs puts it (Sachs 1984, 394; see also Dornbusch 1985; Feldstein 1987). That said, Díaz-Alejandro takes care to emphasize that the debt crisis is also a political crisis for most debtor countries. To the extent that private assets have been built up abroad by capital flight, these 'considerations indicate that the debt crisis is not just a North-South issue; for several Latin American countries it is also an issue of the distribution of domestic income and wealth' (Díaz-Alejandro 1984, 336; see also Lessard and Williamson 1987).

Griffith-Jones and Sunkel

The concluding part of Díaz-Alejandro's analysis is the starting point of the more structuralist analysis of debt proposed by Stephany Griffith-Jones and Osvaldo Sunkel in their book, *Debt and Development Crises in Latin America: the End of an Illusion* (Griffith-Jones and Sunkel 1986). The views developed in this book, which are not always the same as the views developed by Griffith-Jones in her many books and papers (e.g. Griffith-Jones 1988, 1991), are at the radical end of the spectrum of views which are consistent with a system-correction perspective. The nature of the corrections called for by Griffith-Jones and Sunkel are more far-reaching than those suggested by Sachs, Díaz-Alejandro, Dornbusch and others.

Griffith-Jones and Sunkel begin their analysis by noting that a recovery in the global economy will not of itself lead to an ending of the developing countries' debt crisis. In this manner they join with the Brandt Commission in warning against the false optimism built into some debt-cycle models. They suggest that a worldwide economic recovery will be dependent upon the growth of a structurally unsound US economy. The regional multiplier effects of the recovery will also be weak in Latin America because of the dominance exercised by US firms in primary commodity markets. The authors talk pointedly of the prospects for 'immiserating growth' in the Americas.[1]

These observations lead Griffith-Jones and Sunkel to a more fundamental challenge to system-stability thinking on the debt crisis. Griffith-Jones and Sunkel inveigh against the mounting dependency of Latin America's export-led development. They insist that Latin America's economies continue to be at the mercy of unequal global exchanges and the transnational corporations, and that they face an 'overwhelming and implacable necessity to obtain foreign financing' (Griffith-Jones and Sunkel 1986. 26). In short, the nature of Latin America's development has removed from its peoples, and their representatives, the capacity to control their own destiny. It may also have removed from most Latin American economies a form and a structure of development which is able to meet basic needs (see also Prebisch 1982).

A powerful undercurrent to the Griffith-Jones and Sunkel volume insists that the debt crisis will only be solved on the basis of a more radical challenge to the model of development which is being pursued within Latin America. In the 1970s the underlying crisis of development in Latin America was disguised on account of the financial permissiveness of the commercial banks. In the 1980s this glue came unstuck. In future, the governments of Latin America are required to recognize that their 'main responsibility. . .is to their own peoples and not to transnational banks' (Griffith-Jones and Sunkel 1986, 178). In practice, this means a parting of the ways with both import-substitution industrialization and export-led growth. Griffith-Jones and Sunkel call upon the Latin American debtors to escape their financial chains by means of a prior break with the dependent development strategies which they are now enjoined to follow.

Adjustment Policies

System-correction perspectives on the debt crisis do not comprise only the writings of Sachs, Díaz-Alejandro and Griffith-Jones and

Sunkel. To see how this perspective has developed over the years –
and to bring in the views of other system-correction theorists – it
will be helpful to review some system-correction commentaries on
the policing of the debt crisis since 1982.

Efficiency

Consider, first, the question of economic efficiency and how this
issue is dealt with by most system-correction theorists and practitioners. SC theorists point up several problems which beset the
often singular logics which informed the containment and adjustment policies of the 1980s. These logics concern: (1) the assumed
rate of recovery in the global economy; (2) the nature of regional
economic multipliers; (3) the relationships between current account
deficits, external debts and inflation; and (4) questions of
unemployment and investment rates.

(1) Mention of the global economic recovery returns us to the debt
projection models put forward by Cline and the World Bank (see
chapter 3; see also Morgan Guaranty Trust 1983, 1984). The
empirical reliability of these models has been challenged on a
number of occasions. Corbridge has noted that Cline's target figure
of industrial country growth of 3 per cent per annum was just met
for the period 1984–6, as were his assumptions of a declining
LIBOR, a depreciating dollar and a low rate of inflation in the
OECD economies (Corbridge 1988, 116–18). The price of oil also
dropped to below $20 per barrel in 1986, which was well below
Cline's most optimistic prediction, and almost half the price
assumed for 1986 in his base-case model. Despite this, the buoyant
performance of the industrial countries did not sponsor the decline
in LDC debt totals which Cline was forecasting (see table 4.1 and
compare table 3.3). A comparison of the actual and predicted debt
totals for the major debtor countries reveals that debt totals were
continuing to rise into the second half of the 1980s. In the case of
Brazil, total external debt stood at $104 billion in 1984; almost
exactly the figure predicted by Cline for 1986 and 1987. (By 1987
Brazil's total external debt was $124 billion.) Mexico and Argentina
tell a similar story, as do most of the oil-exporting developing
country debtors. Only in Venezuela did the total external debt
stabilize in the manner that Cline suggested it would. Finally, the

continued rise in LDC debt totals (1985–7) occurred notwithstanding the success of most Latin American debtors in making the adjustment efforts required of them. In Mexico and Brazil in 1984 the current account was in surplus above the level predicted by Cline and required by the creditors of the two countries.

Table 4.1   The 'Cline model' evaluated: balance of payments, debt projection and actual totals

	1983	1984	1985	1986
Argentina				
Exports[a]	7910	8017	8396	6852
Imports[b]	4666	4585	3814	4724
Current account	−2439	−2542	−954	−2864
Debt	42 319	45 839	48 444	48 908
Brazil				
Exports[a]	25 127	27 005	25 637	22 396
Imports[b]	16 844	15 209	14 346	15 555
Current account	−6799	53	−273	−4930
Debt	91 145	104 384	106 730	110 675
Mexico				
Exports[a]	21 168	24 054	21 866	16 237
Imports[b]	8201	11 267	13 459	11 997
Current account	5223	3905	+540	−1270
Debt	89 099	97 307	97 429	101 722
Venezuela				
Exports[a]	15 040	13 340	12 272	10 029
Imports[b]	6667	7594	8178	9565
Current account	3707	5298	+3086	−2011
Debt	36 760	34 247	37 079	33 891

Values are in US$ millions.
[a]Merchandise exports.
[b]Merchandise imports.
Source: World Bank (1985b to 1988b)

(2) The optimism of the debt projection models produced by Cline and the World Bank can be explained in a number of ways. One suggestion might be that the performance of the world economy is inherently difficult to forecast. The margins of error found in these models are then not unacceptable. Others would demur, pointing out that this fails to explain why the direction of error is so

consistently tilted in favour of economic optimism. Sam Cole has shown that 'year after year the forecasts of the World Bank lie above the respective trends [suggested by past performance] by several percent' (Cole 1989, 177). Cole suggests that World Bank models of the world economy are calibrated in such a way that they embody the standard assumptions of 'neo-classical economic theory, namely that growth depends on the expansion and improvement of productive facilities, including the division of labour, through international trade' (ibid. 186). The World Bank model is then insensitive to important regional variations in the economic feedback mechanisms it seeks to describe (as might be promoted by selective tariff regimes and by the structure of international commodity markets). The World Bank also assumes a locomotive effect, of industrial countries on developing countries, which its own statistics suggest will not be forthcoming.

Cole proceeds to explain the manner of construction of World Bank economic models in political-cum-sociological terms. He suggests that World Bank forecasts post-1983 have been 'highly conditioned by the dominant political agenda of the US administration, and that this has co-opted the scientific agenda and transformed it into a manipulative exercise' (Cole 1989, 177). The very authority of World Bank forecasts has encouraged Bank officials to substitute 'a narrow selection of results' (ibid. 186) for rigorous scientific modelling – a process which then presents the developing countries with a rosy view of a world economy heading for growth and development. Instead of recognizing an increased instability in the world economy, World Bank models tend to forgo an account of 'restructuring, depreciation and technical change [in favour of] a neat theoretical relationship between growth and capital accumulation' (ibid.). An optimistic quantification rides roughshod over a more pessimistic and qualitative account of a world economy in crisis.

Not all system-correction theorists would subscribe to the subtext of Cole's analysis of World Bank forecasting models. His more general points about linear, theoretical reasoning would, however, be widely accepted. Most system-correction theorists are sceptical of these sorts of abstracted claims and would want to know more about the empirical relationships which describe regional economic linkages in an imperfect world. Some would join Bank officials in noting that even a significant expansion in OECD economic activity might have only a marginal impact on the welfare of the debtor countries if current regimes of protectionism continue. In its *World Development Report* of 1985, the World Bank suggested that 'an increase in protectionism [in industrial countries] big enough to produce a 10 per cent deterioration in the terms of trade of Latin America would cost the region as much as the real interest cost of their entire debt' (World Bank 1985b, 38).

The policy conclusions which may be drawn from this statement are not straightforward. System-stability theorists might see in this statement a vindication of their calls for free trade; some system-correction theorists will insist that such imperfections will not be repealed in the short to medium term and must be taken into account when devising debt management policies for the real world (Laird and Nogués 1989).

(3) A similar question of 'model and/versus reality' marks the debate on the relationship between current account deficits and external debts. It also informs a parallel debate on budget deficits and inflation.

In the case of trading regimes and indebtedness it is too easy to maintain that inwardly oriented economies are always liable to face problems in respect of external indebtedness. Jeffrey Sachs is in agreement with the general proposition that outward-oriented economies perform better in terms of economic growth than inwardly biased economies. He points out, however, that an outward-orientation is not the same thing as a free market trade policy. In the case of South Korea, 'the quintessential open-oriented economy', the state has intervened to provide trade and other incentives to favour exportables. The 'South Korean case belies the simple position often taken by the United States government and the IMF and the World Bank that "small government", as opposed to effective government, is the key to good economic performance' (Sachs 1989b, 16; see also Amsden, 1989). Díaz-Alejandro makes a similar point with respect to the 'formula so popular among staff members of the IMF that "the current account deficit equals the budget deficit"' (Díaz-Alejandro 1984, 365). He notes that a money-balances approach does not hold true for Mexico and Brazil in 1983–4. Díaz-Alejandro also notes that 'the coexistence of a "favorable" current account and very high inflation rates in both Brazil and Mexico in 1983 and early 1984 erodes the credibility of those who argue that elimination of inflation is indispensable for improving the balance of payments' (ibid. 365).

Sachs makes a related but more telling point, as do Cardoso and Dornbusch (Sachs 1989b; Cardoso and Dornbusch 1989). These authors note that the rate of inflation in a country is likely to increase sharply in the wake of containment-adjustment pro-grammes for debt management. The reasons for this have to do with the 'fiscal complexities' which Sachs rightly notes to be underdeveloped in most money-balances models. Simply stated, the cut-off in bank lending to Latin America, combined with local recessions caused by the process of structural adjustment, can lead to a collapse in the prices of non-tradable commodities, and thus to distress for the non-tradables economy. Many firms are then unable to service their dollar-denominated debts. This in turn puts local banking systems in jeopardy, as it did in Argentina, Mexico and Brazil. Governments are then encouraged by the creditor institu-tions to take over the bad loans of the domestic banking system. (Díaz-Alejandro is more forceful. He writes of extreme pressures being exerted upon Latin American governments by the US banks and the IMF to nationalize bad debts: Díaz-Alejandro 1984.) This in turn creates difficulties for the government, which sees its tax base shrinking rapidly even as its debts are increased. Finally, the domestic currency value of the government's external debt rises 'sharply relative to the domestic currency value of the government's tax revenues. . . .In Brazil, for example, what looked like a moderate fiscal burden of the foreign debt suddenly became enormous after the real exchange rate depreciation during 1980–1982' (Sachs 1989b, 21). Inflation may then be just around the corner. The fear of high inflation adds to local pressures by causing nominal interest rates to rise, so further penalizing the government and widening the fiscal deficit. Inflation then takes off.

In its economic survey of Latin America and the Caribbean for 1988, ECLAC notes that 'In addition to hindering growth, the turnaround in net resource flows was one of the factors which fueled the runaway inflation seen by the region in the 1980s, since, in addition to the costs involved in raising the trade surplus in order to finance the outward transfer of resources, the countries also had to finance an internal transfer of funds from the private to the public sector' (ECLAC 1989, 99). ECLAC suggests that 'somewhat over 4% of the 6% change (in terms of GDP) in external resource transfers has been borne by the public sector' (ibid. 99–102). For some idea of what Latin American payments on external public debt would have looked like without absorption of private debt, see figure 4.1.

The point is that to blame all of this on fiscal laxity is facile and unfair; governments can be caught in a web of uncertain economic events and processes which are beyond their control and which are

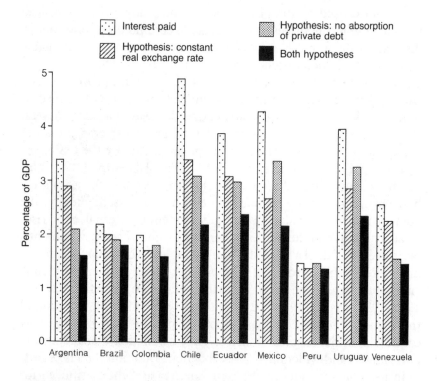

Figure 4.1 Latin America: actual interest payments on external public debt and hypothetical payments at constant real exchange rate and without absorption of private debt, 1982–1987
Source: ECLAC, *Economic Survey of Latin America and the Caribbean*, 1988, 104

bound up with the very policies for structural adjustment they are enjoined to pursue. A long-term process of adjustment can be compromised in the short to medium term by fiscal policies made lax by the need to socialize private sector debts.

(4) A fourth issue concerns the question of unemployment and the levels of economic growth and output within an economy. Marcel and Palma have calculated that in Great Britain alone there was a loss of exports to Latin America equivalent to 49 per cent in real terms between 1980 and 1983 (Marcel and Palma 1988, 389). It has also been estimated that up to one million person-years of employment were lost in North America, between 1982 and 1985, on account of declining exports to indebted countries. This import of unemployment was particularly marked in Florida, Texas and California and has become a political issue in these states. A similar ball-park figure for Europe is one of seven million person years of employment lost between 1980 and 1985 (UNCTAD 1986). Given Keynesian concerns about the economic inefficiency of an unemployment 'equilibrium', this is not a welcome state of affairs. That it is added to by a decline in employment and in per capita incomes in many parts of the indebted world in the 1980s only confirms that the macroeconomic costs of structural adjustment measures have been enormous. Whether they could have been lessened is something we will take up in the fourth section of this chapter.

Morality

System-correction theorists are not only concerned with the economic costs of debt management practices. Although 'economics' remains the rather singular language of this perspective (as compared to the discourse of some system-instability theorists), a related concern is for the moral standing of debt management policies. The Brandt Commissioners express this concern in their suggestion that the debt crisis is a common crisis (Brandt 1983). Precisely because the crisis is not the fault of any one party, and could not reasonably have been predicted, it is unreasonable that the burdens of the debt crisis should be carried by one set of actors more than by another.

The Commissioners also suggest that it is unwise and perhaps immoral to police the debt crisis in such a way that a crisis of international banking is made good at the expense of the needs of people in developing countries. This view can be generalized by means of a Rawlsian theory of justice (see also Cornia et al. 1987; Taylor 1987). In opposition to the just deserts arguments advanced by some system-stability theorists, this argument would call attention to two arguments of equal, if not greater moral force: namely, (a) is it reasonable, by virtue of structural adjustment, to condemn large numbers of people to a life devoid of some basic human needs; and (b) is it proper that the burdens of structural adjustment should fall most violently, in absolute terms, upon the poor and powerless in an indebted country, when they were not signatories to, or major beneficiaries from, the 'profligacy' made possible by international credit money transfers? In Rawlsian terms, the answer to each question must be 'No, it is neither fair nor reasonable'. The application of an overriding moral principle – in this case the principle of moral hazard – here works to condemn fellow humans to a form of existence which few rational outsiders (those living under Rawls's veil of ignorance) would be prepared to accept. It also fails to distinguish between the active and passive progenitors of the policies in which moral hazard is thought to reside.

Politics

Finally, there is the question of the politics of structural adjustment. We have seen that system-correction theorists call attention to the political and normative contexts in which lending, borrowing and debt-servicing decisions are made. Structuralists, in particular, are wont to emphasize the manner in which debts were contracted by the elites of many problem debtor countries to further programmes of development which were of little relevance to the needs of ordinary citizens. A not dissimilar point has been made with respect to the politics of structural adjustment programmes. System-correction theorists are sensitive to the political contexts within which 'poor' policy decisions can be made.

This is not a question of referring back all decisions to the poor qualities of a rent-seeking society or regime. The political context tends to be much more finely worked out, with attempts being made to differentiate between regime types (Haggard and Kaufman 1989; Nelson 1990) and to situate regime types in relation to local structures of class and civil society (Frieden 1987a). The very possibility of structural adjustment programmes being put into effect now becomes an object of study, as does the likely manner of their operation (given various degrees of political support and opposition). Again, this brings us back to questions of timing and of political and economic trade-offs. Jeffrey Sachs sums up the position of many system-correction theorists when he calls for a financial time-out – for a period of grace during which debtor countries and their governments can find their feet and lay the foundations of a more stable economic growth. This will be a time during which some hard decisions can be taken, unpressured by immediate debt servicing demands and difficulties. What Sachs and others are wary of is not the idea of structural adjustment, but of high-handed IMF conditionality programmes to enforce it come what may. IMF conditionality then becomes a point of focus for oppositional movements. Structural adjustment in this manner is likely to be politically destabilizing, at least in the short term. It is likely to make the option of repudiation more likely rather than less likely (Rodrik 1990). To provoke such an action is to act in a manner which is unpragmatic.

5 Policies for Debt Crisis Management

System-correction policies for the management of the developing countries' debt crisis vary widely in intention and design. We will outline several such proposals in this section. Before that it is worth noting that system-correction theorists are in broad agreement on two points: first, that there is an historical precedent for burden sharing; and second, that 'debt service is ultimately best guaranteed by investment and growth' (Cardoso and Dornbusch 1989, 133).

Not all economists have been willing to examine the historical record of debt crisis management. Among those who have, the work of Eichengreen and Lindert stands out – in particular, a collection of essays edited by these two authors entitled *The International Debt Crisis in Historical Perspective* (Eichengreen and Lindert 1989a). Eichengreen and Lindert are sensitive to the dangers of making false comparisons between the debt crises of the 1890s and 1930s and that of the 1980s. Unhelpful lessons can easily be learned. Nevertheless, their researches clearly suggest that 'previous debt crises have usually ended in some forgiveness. A compromise is typically reached in which the debtors service some, but not all, of the debt that is due. *A partial writedown of the debt is the norm, not the exception*' (as summarized by Sachs 1989b, 23; emphasis in the original).

Eichengreen and Lindert point out that 'governments have always been intimately involved in the process of debt readjustment' (Eichengreen and Lindert 1989b, 7). This was the case in the 1930s and 1940s, notwithstanding the fact that most external debts were then contracted in the bond markets. Eichengreen and Lindert note that British investors in the 1930s were represented by the Corporation of Foreign Bondholders, an organization which maintained the closest of links with the British government. The main difference between the 1930s and the 1980s is not with regard to the fact of government intervention in international financial matters, but rather with respect to its direction and scope. Thus, 'In the 1980s creditor-country governments, motivated by the desire to protect their banking systems, have exerted the greatest pressure on the debtor countries, urging full repayment and macroeconomic adjustment. . . .In earlier periods creditor-country pressure operated in more ambiguous directions. Sometimes governments viewed the obstinance of private creditors as an obstacle to the cultivation of harmonious international relations' (Eichengreen and Lindert 1989b, 7). Eichengreen, Lindert and colleagues also make the point that the practice of 'buying back defaulted bonds in the markets as a way of liquidating a debt overhang' (ibid. 9) is not an invention of the 1980s; it was common enough in the 1930s and 1940s.

Finally, the researches of Eichengreen, Lindert and others make it clear that moral hazard is not unambiguously increased by a debt management strategy which is based on partial debt forgiveness. One of the reasons for this is that bankers have short memories. Thus, 'In the early 1970s it was not uncommon for bankers to assert that sovereign default was inconceivable' (Eichengreen and Lindert 1989b, 4). notwithstanding the events of the 1930s and 1940s. In any case, banks tend to look to the future, seeking to determine creditworthiness now and in the years ahead. The fact of the matter is that from 1940 to 1970 Brazil had no more difficulty in borrowing from private sources than did Argentina; yet Argentina repaid its debts in full in the 1930s and 1940s, when Brazil did not. If history does have a lesson for us, it is not just that default is recurrent, but that 'the countries that default tend to be the same ones, generation after generation' (Eichengreen and Lindert 1989b, 4). The pragmatic conclusion to be drawn from this is that abstracted notions of moral hazard can be exaggerated. A bit of give and take with respect to the repayment of debt in arrears is not a major threat to the integrity of a market-based economic system.

A second point common to most system-correction theorists concerns the economic climate in which structural adjustment policies are best pursued. This point has two dimensions to it. On the one hand, it is suggested that problem debtors cannot be expected to meet their debt service obligations unless the international economic environment is improved. This will entail greater ease of access to industrial country markets, lower real interest rates and sustained economic growth in the OECD countries. This point was made very firmly by the Brandt Commission in its report on the debt crisis as a common crisis (Brandt 1983; see also Lever and Huhne 1987). It was also developed in various debt prediction models which suggested that debt/export ratios would not decline except on the basis of a moderate-to-good recovery in the OECD world economy (see figure 4.2). The Brandt Commission also maintained that problem debtor nations faced a serious crisis of solvency and that new funds should be made available to them. Specifically, the Brandt Commission proposed that IMF quotas should be at least doubled and that there should be a major new allocation of Special Drawing Rights ($10–12 billion per annum for at least three years in the early to mid-1980s). The Commission also proposed the setting up of a World Development Fund to sit alongside the IMF and the World Bank. The new Fund would be charged with many of the rights and responsibilities which Keynes had proposed for the Bretton Woods institutions at the time of their founding in 1944.

A second dimension to this commentary has regard to the situation within the problem debtor countries. Most system-correction theorists are in favour of structural adjustment policies

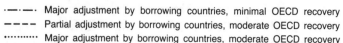

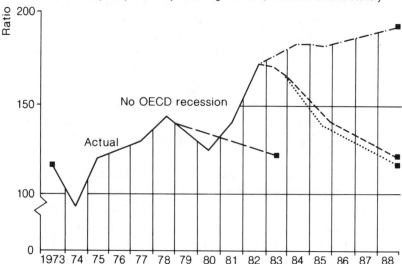

Figure 4.2 The debt/export ratios of 21 major developing
country borrowers (average of beginning and
end-year total debt as a percentage of exports of
goods and services)
Source: Morgan Guaranty, after Frances Williams

of one sort or another, but they are wary of the suggestion that from
the ashes will arise a new phoenix (Edwards 1989). A persistent
weakness identified in the containment strategy is its willingness to
assume that an economy in debt servicing difficulties can grow
quickly on the basis of a severe pruning of its assets. This argument
seems to neglect the supply side of an economy as much as it
neglects local demand conditions. A pruning argument might hold
in the abstract and over a longish period of time, but it seems to
skirt over some critical questions. For example, if resources are to
be earmarked for debt service above all else, from where is
investment income to derive? If firms close in the wake of a
recession, from where are new exports to derive? (The answer to
this last question depends upon such things as the rate of new firm
formation and the efficiency of expenditure switching within an

economy). And even if debt service payments can be made, where does this leave development?

Again, it is a matter of timing and priorities. The system-correction perspective is concerned that economic and political affairs should not be so disturbed in the short run that a long-run process of structural adjustment is put in jeopardy. This will happen if there is a 'premature transfer of real resources to the creditors' (Cardoso and Dornbusch 1989, 132), if future investment funds are put at risk, and if debtor countries are pushed in the direction of default and repudiation. System-correction theorists also maintain that there is an alternative to the containment strategy. In general terms, the alternative is to act pragmatically and to seek an equitable sharing of the burdens of adjustment. More straightforwardly, it is to buy time now in order to allow future debt repayments to be serviced from a sustainable local growth and development. (This process of buying time may involve greater government regulation of financial markets and a willingness to increase official development assistance. Greater definition of the role of the lender of last resort would also be helpful: Guttentag and Herring 1985; Dale 1985.)

In more detail, system-correction strategies range from the sort of wide-ranging prospectus set out by the Brandt Commission to myriad smaller strategies, most of which were proposed to deal with specific aspects of the banking debt crisis. Most of these strategies were not acted upon and therefore it is difficult to assess what their likely effects would have been. In the present context it will suffice to indicate a few of the more widely discussed strategies which were put forward in the first and second halves of the 1980s. A first group of proposals are from the early 1980s. They include:

1   A scheme proposed by Felix Rohatyn, wherein 'bank claims on developing countries would be converted into long-term and low interest bond issued by an existing or newly created international agency' (Nunnenkamp 1986, 171–2; Rohatyn 1983). This agency would provide debt relief to developing country borrowers by making available long-term loans at concessional interest rates. Rohatyn assumes that governments in the OECD countries will bail out those of their commercial banks with liquidity or solvency problems. Rohatyn's scheme is not dissimilar to a scheme proposed

by Peter Kenen, also in 1983. Kenen argued that bank claims on developing countries should be purchased by an International Debt Discount Corporation at a 10 per cent discount (Kenen 1983). The Corporation would then use some of the discount to grant debt relief, while also extending the maturities on new loans made to developing countries.

2   A second set of schemes involves proposals to cap interest rates on loans. The intention here is to limit LDC debt service obligations to a fixed proportion of local export earnings. The idea of a cap is not at odds with the schemes proposed by Rohatyn and Kenen. It was mooted, in particular, by eleven ministers of Latin American countries who were signatories to the Consensus of Cartegena in June 1984. It was also put forward, in another form, by Bailey in *Business Week* (Bailey 1983; for a review of such schemes, see Bird 1987).

3   A third set of schemes proposed a system of insurance for the portfolios of commercial banks. As developed by Wallich (1984), this scheme would aim to call forth new bank lending to the debtor countries on the basis of a prior reduction in the risks facing banks engaged in this enterprise. The scheme would be financed by the banks themselves and by creditor-country governments. As envisaged by Wallich, the scheme would not diminish the level of debt outstanding. Wallich's proposal is also consistent with a range of schemes to improve the regulatory regimes which surround commercial bank lending. Some of these proposals were acted on in the later 1980s.

It is a sign of the times that William Cline referred to all of the schemes just mentioned as 'proposals for radical reform' (Cline 1984, 130–3). In fact, most of the schemes are concerned primarily for the safety of US banks. Schemes from the second half of the 1980s include:

1   The Bradley proposals (first mooted in 1983, but seen very much as an alternative to the Baker proposals of 1985–7: Bradley 1986). Bill Bradley, a Senator from New Jersey, proposed that debt relief should be offered to some middle-income highly indebted countries. The countries would be in Latin America, where Bradley feared a further loss of US export markets. Up to 30 per cent of the

relief would be earmarked for Brazil. The debt relief 'would be financed by a tax on the commercial banks and channelled through the multilateral agencies' (Edwards 1988, 22). An early version of the Bradley proposals suggested that debt relief would mark down the value of specified loans by 3 per cent. Bradley also suggested that interest rates should be cut by 3 per cent for a period of one year. This would provide debtor countries with the time needed to sort out their domestic economic affairs. Significantly, in 1986 the Bradley proposals were denounced by Paul Volcker, for being 'entirely unrealistic, because they involved a loss of 6–7% on banks' Latin American assets, or US$12 billion spread over the 24 US [banking] majors' (Roddick 1988, 229).

2   Proposals to link partial debt write-downs to guarantees in respect of reforms in international trading markets and in domestic economic affairs. Simple versions of this proposal would take the form of compensatory financing facilities and schemes to manage risk in commodity markets (such as the World Bank introduced in 1989). More far-reaching proposals include the one put forward by Cardoso and Dornbusch for Brazil. These authors propose a 'scheme that recycles a large part of the interest payments in the country. . .[thus doing] away with the need for trade supluses and the resulting crowding out of investment' (Cardoso and Dornbusch 1989, 133). Debts would be repaid by means of investment certificates, the proceeds from which could not be transferred out of Brazil.

At this point we should stop and take stock. The proposals outlined above are only a small sample of the more than seventy proposals put forward in the 1980s to deal with the question of debt crisis management. It will be apparent that we have not yet considered more radical proposals for extensive debt forgiveness, such as the Brandt Commission proposed with respect to the official debts of many African countries. Some of these proposals will be discussed in the next chapter, although system-correction theorists of a structuralist persuasion will not be averse to some of them. By the same token, the proposals we have outlined are not exclusive to a system-correction perspective, nor have all of them been ignored by the World Bank and other agencies involved in practical debt crisis management (see chapter 6). Finally, these proposals are not intended to substitute for more long-lasting adjustment program-mes. Jeffrey Sachs takes a typically pragmatic line when he calls for changes in the regulatory regimes which surround US commercial banks, for a financial time-out to be paid for by debt relief, for domestic economic reforms, and for reforms of the international financial and trading systems (Sachs and Huizinga 1987; Sachs 1989b). The idea that partial debt forgiveness should also be related to good behaviour is not discounted, although most system-correction theorists would reject attempts to privilege this relation-ship over all others (see Corden 1988). In this respect they are critical of IMF conditionality programmes and of the tendency of some Fund officers to interpret the world in terms of abstracted economic models (Carmichael 1989).

6 Conclusion and Critique

It is worth reiterating that the system-correction perspective defines the middle ground of debt and development studies; as such it is home to many more scholars than the system-stability or system-instability perspectives. Although the differences and debates apparent within the system-correction perspective are less exciting than the divisions between the three competing paradigms, this is where most academic discussion is centred. Scholars working in this area might reasonably be described as Keynesians in terms of the forms of economic reasoning to which they often subscribe, even if a penchant for *dirigisme* is no longer (if indeed it ever was) a point of principle for this perspective.

Looked at from the outside, the system-correction perspective is a moving target, as befits its attachments to Keynesianism and to pragmatism. The criticisms which can be put to it are not hard to anticipate. From the system-stability perspective comes the charge that moral hazard is endemic in system-correction proposals for debt crisis management, and that an unhelpful tinkering with imperfect economic systems is no substitute for more fundamental and pro-market economic reforms. From the system-instability perspective comes the charge of reformism. SC theorists are accused of a toothless liberalism and of failing to recognize that the debt crisis is symptomatic of more chronic instabilities in financial capitalism. It is to a discussion of these 'chronic instabilities' that we now turn.

Note

1   This phrase was popularized by Jagdish Bhagwati in the 1960s.

FIVE

The Debt Crisis: a System-instability Perspective

1 Introduction

A third perspective on the developing countries' debt crisis overlaps in part with a radical Keynesian perspective, but in most respects is quite singular. The system-instability perspective refuses to examine the debt crisis except in terms of a wider analysis of certain crisis tendencies within the global political economy. The debt crisis is thus symptomatic of a deeper malaise. Although a simple base and superstructure model is no longer accepted on the left, radical political economy adheres to an epistemology which seeks to relate empirical phenomena to deeper structures or systems within a social formation. In this case, the debt crisis is one moment in a crisis-ridden transition from Fordism to post- Fordism, in a crisis of globalization written through fictitious capitals, and/or in a crisis of developmentalism wherein profits are put before people. The geographical incidence of the debt crisis in turn is expressive of a geopolitical battle which is being waged to avoid the costs of devaluation that are linked to the deeper crisis of capitalism. This battle is fought out within indebted developing countries, between creditors and debtors, and between the USA and its rivals. The battle is also fought out with respect to the Bretton Woods institutions, and with regard to the 'other' debt crises which mark the USA (especially its farming regions), Australia, Eastern Europe and the ex-Soviet Union.

Although it is possible to itemize in this way the main features of a system-instability perspective on debt, it will be apparent that

system-instability accounts of the developing countries' debt crisis
vary markedly in terms of the processes of dislocation which they
choose to emphasize. One aim of this chapter is to reflect this
variety, rather than to seek to disguise it. Radical political economy
is an umbrella term and this needs to be remembered. In order to
meet this and other objectives, the chapter is organized as follows.

Section 2 outlines some of the economic and political claims
which underpin most system-instability accounts of the debt crisis.
The section begins by reviewing the basic propositions of Marxist
political economy. It also considers how a diverse Marxism has
been combined with more populist perspectives to create a broad-
based radical political economy. Some of the moral and political
outlooks associated with this perspective are outlined.

Section 3 examines the relationships between radical political
economy and what might be called radical development studies.
This entails a brief survey of neo-Marxist theories of development,
of *dependencia* theories and of more recent attempts to theorize the
internationalization of capital. It goes without saying that much will
be left out of this brief review. (A guide to further reading is
implicit in the text, as it is also in chapters 3 and 4.) It should also
go without saying that the labels attached to some of these
intellectual traditions (for example, neo-Marxism) are problematic;
as ever, the reader is encouraged to look beyond the label and to
examine the substance of what is being said.

Section 4 considers how a system-instability perspective on the
debt crisis mobilizes certain of the ideas and claims set out in
sections 2 and 3. Particular attention is paid to the notion of
critique. System-instability accounts of the debt crisis are notable
for the way in which they proceed by first contesting the views
expressed in competing intellectual traditions. Attention is also
directed to some of the accounts which have been put forward to
explain the alleged chronic instabilities of modern capitalism.
These accounts can be more or less populistic (as in the work of
Payer and George on *Ponzi Schemes* and the *Money Mongers*), and
more or less academic (as in the work of Harvey, Lipietz and others
on the crisis tendencies endemic to particular regimes of accumulation within capitalism). References to a populistic approach are not
made with a derogatory intent. Although there are problems with
some aspects of the radical populist approach, an attempt to

popularize its main claims is not one of them. Finally in this section, there is an account of the geopolitics of debt. The focus of this account is on the United States and its powers with respect to international monetary circuits and institutions (including the IMF). The main guide to the issues raised in this sub-section is Riccardo Parboni.

Section 5 is concerned with system-instability proposals for an end to the debt crisis. These proposals do not generally take the form of proposals for debt crisis management, in the sense of a proto-Keynesian system of economic and political bargaining. The proposals consist rather of calls for collective debt repudiations by developing countries, and/or of calls for debt to be written down as part of a democratically organized and 'people-centred' process of development. Susan George calls this last solution the 3-D solution: Debt, Development and Democracy (George 1989). These proposals also express the moral concerns of most system-instability theorists. Although it would be facile to maintain that system-instability theorists have a monopoly on fellow-feeling for the victims of the debt crisis in the 'Third World' (see chapter 6), the system-instability perspective does address the plight of these people more directly than does the system-stability perspective or the system-correction perspective. The system-instability perspective is much less tied than its rivals to 'economics' as the language in and through which its narratives are written. The chapter concludes with a critique of the system-instability perspective on the debt crisis (section 6).

2 Marxism and Radical Political Economy

Trying to write a brief introduction to Marxist political economy is a thankless task. More so even than Keynes, Marx was a visionary thinker whose work encompassed history, philosophy and politics as much as economics; indeed for Marx and most of his followers these disciplinary boundaries do not count for much. Marxism is also different from the diverse works of Marx himself, just as Keynesianism is different from the works of Keynes. The Marxist tradition has been extended, appropriated and bastardized by generations of Marxists, neo-Marxists, structural Marxists and post-Marxists. More recently, the death of Marxism has been proclaimed along with the death of socialism and the end of history (Fukuyama 1989). In this chapter we will have little time for such 'endism'. Even if Marxism is deficient as a guide to politics, it need not follow that Marxism is redundant as a guide to the changing contours of capitalism and modernity.

Class and Exploitation

These points made, let us review those aspects of Marxist economics which have most bearing on system-instability accounts of the debt crisis. An obvious starting point is the concept of class and the Marxist theory of exploitation. Class is important to Marxian economics in a philosophical sense as well as in a technical sense. Class is important in a philosophical sense because it confirms that Marxism is opposed to the tenets of methodological individualism.[1] According to Marx, economic actors are always the bearers of social relations; that is to say, an economic actor is able to act only with regard to certain limitations imposed by his or her class location, gender, ethnicity and so on. In terms of class, individuals in capitalist societies are capitalists or rentiers or labourers and their subjective preferences derive in part from these wider points of reference. Marxism subscribes to a transformative or relational account of human nature, and not to the essentializing anthropology to be found in neo-classical economics (Ollman 1976).

The technical bases of class and exploitation are related to these philosophical concerns. Marxism suggests that the history of pre-socialist societies is bound up with a history of struggle around and between the forces of production in a given society and the social relations of production which determine the access of individuals to the means of production. In feudal societies the main contending classes are landlords and serfs, or landlords and peasants. In capitalist societies the main contending classes are the bourgeoisie and the proletariat (which is not to say that a landlord class has disappeared, or that there are no divisions, or fractions, within these grand classes). By the same token, in a feudal society the exploitation of the peasantry is organized by means of political institutions and with the use of force. In capitalism, the exploitation of the working class is organized through economic relationships which are at once more efficient and less transparent than their counterparts under feudalism. Moreover, this exploitation is so ordered that it makes necessary a constant revolutionizing of the productive technologies of capitalism. This is very different to the situation under feudalism and it prompts Marx to pay tribute to the revolutionary qualities of capitalism. Although Marx was sensitive to the contradictions of capitalist accumulation, he saw that capitalism gave rise to that combination of pleasures and pains which defines the modern condition. When Marx and Engels proclaimed that, under capitalism, 'all that is solid melts into air, all that it is holy is profaned' (Marx and Engels 1967, 83), they offered their remarks as a paean to change as much as a complaint about its consequences.

The Labour Theory of Value

At this point a little detail is unavoidable. Consider the Marxist account of class and exploitation as it pertains to capitalism. Marx suggests that capitalism is born of a process of primitive capitalist accumulation wherein a majority of the population are created as free wage labourers divorced from the means of production (Marx, 1976). The means of production are concentrated in the hands of a new class of capitalists (and their allies in the countryside). At this point the nature of exploitation begins to change. Marxism defines exploitation as unpaid labour. Within a feudal economy the amount of unpaid labour performed may be considered to be a matter of simple calculation. If a person works on his or her land for three days a week, and then performs unpaid labour for his or her landlord for three days, we might speak of a rate of exploitation of 3/6 (or 50 per cent). But what of the labourer in a capitalist system? The person who puts in a ten or twelve hour shift in a nineteenth century mill is paid a daily or a weekly wage for this labour. The pay-slip does not say 'Four days for the worker, two days for the boss'. Yet conditions in many towns in nineteenth century England were poor for most freely employed and 'fairly rewarded' labourers. At the same time the income gap between rich and poor seemed to be getting wider. (Here we tilt towards more popular renditions of Marxism and thus towards a broader tradition of radical political economy.) How can this be and wherein lies the exploitation to which Marx refers?

The answer lies within the working day and at the point of production. Marx suggests that all commodities are exchanged according to 'the amount of labour time required to produce them under the conditions normally obtaining' (Smith 1986, 285). A chair which takes two hours to make will exchange for half a table, if a table takes, on average, four hours to make. The same holds true of labour when it is in the form of a commodity. The wage which labour-power commands is 'determined by the socially necessary labour required for subsistence – the cost of production (and reproduction) of labour itself' (ibid.). The twist comes with Marx's contention that the exchange value of labour (labour-power) may be exceeded by its value to the capitalist who purchases it as a commodity. A ten hour shift by a labourer may then consist of six hours work for the labourer (producing enough commodities to pay for his or her subsistence) and four hours unpaid labour for the capitalist. This unpaid labour is the source of the surplus value which accrues to the capitalist and which forms the basis of his or her profit. As Smith points out, however, 'It is ownership of the means of production that enables the capitalist to engage in exploitation, whereby part of the product of labour is appropriated by the capitalist. Value and class relations are thus inextricable elements of the social practice of production under capitalism' (Smith 1986, 285).

What we have just outlined is the simple theory of exploitation as it is set out by Marx. It should be clear that, for Marx, labour is the source of all value – a common enough proposition in nineteenth-century political economy. If this seems odd it can be explained by reference to the concept of capital in Marxism. Marx defines direct living labour as variable capital (v). Dead, or past embodied, labour – labour that has been expended to produce machines, or capital in a conventional sense – is then called constant capital (c). Between them, constant and variable capital (or labour) are the source of all value.

On the basis of this simple model, Marx derives some equally straightforward equations. Thus, if surplus value is S, the total value of commodity (Y) is given by: $Y = C + V + S$. The rate of surplus value, or the rate of exploitation (r) can be written as $r = S/V$. This is not the same as the rate of profit, which is of more concern to a capitalist. The rate of profit (p) has to take into account expenditure on constant capital and is defined as $p = S/(C + V)$. Finally, a hint as to the importance of constant capital in Marxian accounts of accumulation and crisis is given by the identity of the organic composition of capital (q). This is written as: $q = C/(C + V)$. The reader might note that an increase in q will be associated with a fall in p if the rate of exploitation is unchanged (this section is after Smith 1986).

Accumulation and Contradiction

Thus far we have outlined only the most static elements of Marxist economics. Critics and proponents of Marxism would agree that what makes Marxism plausible is its sensitivity to the dynamics of accumulation and crisis formation within the capitalist mode of production. Marx's account of a capitalism which is in a state of permanent revolution is a visionary account by the standards of mid-nineteenth-century social thought. His sense of the contradictions inherent in the modern, capitalist condition anticipates the later observations of his fellow Germans, Weber and Simmel. The big difference between these thinkers is that Marx shares the optimism of his age. Marx believes that the contradictions inherent in capitalism can be transcended by means of a more rational organization of society (Marx 1974). This forms the basis of his account of socialism-communism. Weber and Simmel, writing some fifty years later, are less inclined to trust to reason to overcome the contradictions inherent in all forms of modern life, including socialism (Weber 1978; Simmel 1990; see also Sayer 1991).

It would be easy to lose sight of our story at this point. If we return to the more narrow concerns of Marxist economics, we see that Marx was able to move on from his theory of exploitation to fashion an account of the processes of accumulation under capitalism. His first observation is that the labour relation describes a process of class struggle about the point of production. Because surplus value and profits are defined as residual categories, it follows that they can only be increased at the expense of the working class. In relative terms, at least, the Marxian model is a zero-sum model.[2] The capitalist is interested in making the labourer work a longer day and/or more productively; the labourer is concerned to minimize the unpaid labour which he or she must surrender.

This simple observation is pregnant with others besides. Consider the labour relation from the point of view of the capitalist. Marx maintains that the capitalist class is interested in increasing the absolute surplus value which it extracts from the working class. It can do this by insisting on longer working hours, by introducing shift-work and so on. There are limits to this as a strategy, however. Absolute surplus value can only be expanded to the limits of the 24-hour day and it may be reduced by successful collective actions organized by the working class. A more realistic option open to the capitalist is to increase relative surplus value by introducing new machines and technologies which compel the worker to work more productively. An individual capitalist is under further pressure to pursue this option because of the competitive pressures which he or she faces from other capitalists. A firm which is not innovative will go to the wall – an observation on which all students of capitalism are agreed. Innovation is not incidental to capitalism, as it perhaps was to feudalism. Marx adapts an aphorism of Plato's to argue that, under the rule of capital, necessity is the mother of invention. Capitalism is inherently dynamic; its landscapes and its production technologies, even its cultural and political relations, are in a continual state of flux.

This is not the place to follow this logic sideways into Marxist theories of the state and civil society. It will be apparent that the state, in this schema, is required to play a role whereby it secures the conditions of existence of capitalist accumulation (Jessop 1982). The state will act to secure property rights and the reproduction of labour. Our focus must instead be on the contradictions engendered by the processes of accumulation we have just described. Briefly stated, these contradictions are three-fold. First, the process of production itself will be problematic to the extent that the working class contests the longer hours and lower real wages which are periodically demanded of it. Second, attempts to increase profits by driving down the real wages of labourers may be counter-productive. Products must not only be produced at a profit, they must also be sold or realized. A capitalist puts money capital into the process of production in the expectation that his or her money will 'grow' as a result. A working class unable to consume is not advantageous to capital in its attempt to close this particular circle (as Keynes was to note much later).[3] Finally, the logic of capitalist accumulation is contradictory because it depends upon the progressive substitution of constant capital for variable capital. This process of innovation may be profitable from the point of view of an individual firm, but to the extent that the innovation is widely imitated a rise in (q) may force a general fall in (p). This is referred to as the tendency of the rate of profit to decline. Although it can be offset by an increase in the rate of surplus value, it may be viewed as symptomatic of a deeper crisis of over-accumulation in capitalism.

Crisis Formation and Displacement

This last point is in need of some amplification. What we have just outlined is a simple reproduction schema in Marxist economics (see figure 5.1). It is the sort of schema which encouraged some Marxists to believe that the demise of capitalism was inevitable and that socialism would replace it as a matter of course. Over the past seventy years this prospect has become less and less likely, especially in the industrial economies. Marxists have endeavoured to explain why this should be the case. Not surprisingly, their explanations vary greatly and we cannot cover all of the relevant contributions in this chapter. Attention will be drawn instead to three particular developments in twentieth-century Marxist and Marxissant thought: to the theory of imperialism; to work on crisis formation and displacement in the primary, secondary and tertiary circuits of capitalism; and to work on regimes of accumulation and the regulation of capitalism. Each of these theories develops insights which are of relevance to the system-instability perspective on the debt crisis. Each one builds, too, upon the most basic points outlined thus far: that capitalism is class divided, that it is exploitative and that the process of accumulation in capitalism is inherently unstable and contradictory.

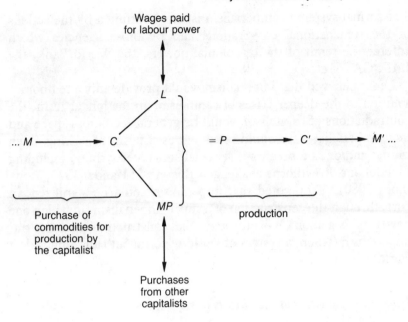

Figure 5.1   Marxian economics: a simple reproduction schema

Imperialism

One response to the crisis of competitive capitalism is imperialism. Lenin adapted the work of Marx and Hobson to argue that imperialism, or monopoly capitalism, is the highest stage of capitalism (Lenin 1970). Imperialism emerges by means of processes of concentration and centralization that are inherent in competitive capitalism. At this time monopoly banking capital merges with monopoly industrial capital to form finance capital. Finance capital is associated with the export of capital from one region to another. This in turn is associated with the formation of international monopolies which divide the world among themselves. In other words, a prior and essential class division in capitalism gives rise to a conflict between territorial powers for the control of space. The periphery of the world system becomes an outlet for the contradictions of metropolitan monopoly capitalism.

The formal system of imperialism is later continued by the actions of the transnational corporations and by those agencies which enforce the remit of the ex-colonial powers (the World Bank, the IMF, GATT etc.).

Lenin believed that imperialism would provide only a temporary solution to the deeper crises of capitalism. In the longer term, the contradictions of capitalism would be generalized across space and inter-imperialist wars would break out as a means of writing down capital and/or as a precursor to socialism. Later writers, including Mandel and Rowthorn, challenged this view (Mandel 1975; Rowthorn 1980). They noted that imperialism could be stabilized, in part, through the construction of a super-imperialism (in which one power is hegemonic) or through the construction of an ultra-imperialism (where a system of world government runs alongside a world economy).

Time, space, crises and the circuits of capital

Lenin's work on imperialism as a spatial fix for capitalism has been refined and generalized by a number of scholars, among the most notable of whom is the geographer, David Harvey. In a lengthy and sophisticated analysis of *The Limits to Capital*, Harvey (1982) paints a picture of an unstable capitalism veering always to crises of one sort or another, only to be returned to an unstable 'equilibrium' by a painful process of economic, political, cultural and geographical restructuring. Time and space are central to Harvey's account of crisis formation and displacement under the rule of capital. Harvey argues that contradictions in the primary, or productive, circuit of capital can be dampened down by the export of capital to its secondary and tertiary circuits (see figure 5.2). The crises then reappear in new forms.

The secondary circuit of capital comprises capital invested in the built environments for production and consumption. Capital flows into this circuit when there is over-accumulation in the primary circuit of capital. Harvey points out that a 'flow of capital into the secondary circuit [presupposes]. . .the existence of a functioning capital market and, perhaps, a state willing to finance and guarantee long-term, large-scale projects with respect to the creation of the built environment' (Harvey 1989, 65). In this manner, Harvey

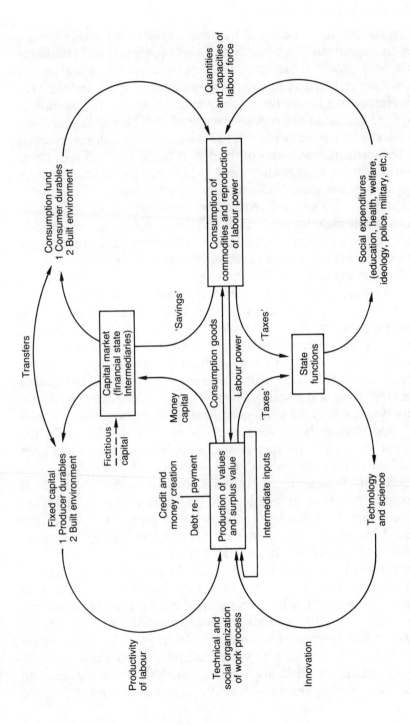

Figure 5.2 The structure of relations between the primary, secondary and tertiary circuits of capital
Source: Harvey (1989, 67)

suggests that the creation of fictitious capitals – or credit monies which are extended in advance of actual production and consumption – is vital to the extended reproduction of capital in its secondary and tertiary circuits. The conquest of space and time are for Harvey two means by which an initial tendency for the rate of profit to decline in one region or sector of an economy can be offset by development elsewhere. This suggestion is central to most system-instability accounts of the debt crisis, as we shall see. Third World debt amounts to an attempt to buy time for capitalism (quite literally) in new spaces. The tertiary circuit of capital offers similar possibilities to those offered by the secondary circuit of capital. This circuit includes investment in science and technology and 'a wide range of social expenditures that relate primarily to the processes of reproduction of labour-power' (Harvey 1989, 66).

These categories allow Harvey to build a sophisticated account of the different crises which must rack modern capitalism. Among these crises are sectoral switching crises (where fixed capital formation is switched to another sphere, such as education), geographical switching crises (which involve the production of new spatial configurations at all scales from the urban to the international) and global crises (which generalize these crises at the level of the capitalist world system). Harvey suggests that 'there have been only two global crises within the totality of the capitalist system – the first during the 1930s and its World War II aftermath; the second, that which became evident after 1973 but which had been building steadily throughout the 1960s' (Harvey 1989, 71).

Another dimension to Harvey's work concerns the idea of struggle, and thus of politics and geopolitics. Harvey does not believe that the crisis tendencies within a capitalist economy are played out in an orderly manner. The precise configurations of crisis – in terms of their incidence in time and space – are determined by the activities of contending social groups. This is true at the workplace, as we have seen, and it is also true with respect to the politics of economic restructuring at the urban and regional scales (the politics of place competition), and at the national and international scales. In the case of the global crises which face capitalism, it is significant, Harvey suggests, that the configurations of crisis are written through the institutions and circuits of money; hence the post-1973 crises of inflation, debts and deficits. The resolution of these crises in turn is bound up with geopolitics and with struggles to avoid the costs of devaluation. Crises, says Harvey, 'unravel as rival states, possessed of different money systems, compete with each other over who is to bear the brunt of devaluation. The struggle to export inflation, unemployment, idle productive capacity, excess commodities, etc., becomes the pivot of national policy. The costs of crises are spread differentially according to the financial, economic, political and military power of rival states' (Harvey 1982, 329).

Crisis and regulation

A third set of propositions on the crises of capitalism derives from a body of work known as the theory (or theories) of regulation. Regulation scholars, including Aglietta and Lipietz, have found it useful to combine aspects of Marxian and post-Keynesian economics with various theories of institutions. Their hybrid theories are then used to provide a periodization of capitalism and of the changing nature of its patterns of accumulation, regulation and crisis formation. To this end, the regulation school has advanced two main concepts. A *regime of accumulation* 'describes the fairly long-term stabilization of the allocation of social production between consumption and accumulation. . .[both] within a national economic and social formation and between the social and economic formation under consideration and its "outside world" ' (Lipietz 1987, 14). A *mode of regulation* describes a 'set of internalized rules and social procedures' which ensure the unity of a given regime of accumulation and which 'guarantee that its agents conform more or less to the schema of reproduction in their day-to-day behaviour and struggles' (Lipietz 1987, 14–15).

Regulation scholars have suggested that the global crises of the 1930s and 1940s and the 1970s and 1980s correspond to two periods of transition between established regimes of accumulation and their associated modes of regulation. In the 1930s and 1940s, a pre-Fordist, or exclusionary, regime of accumulation gave way in the industrialized world to a Fordist or inclusionary regime. Fordism is inclusionary because the working classes are included in the circuits of capitalist consumption as well as production. Under Fordism a system of mass production is complemented by a system of mass consumption wherein the working class is able to consume some of the products of its labours by means of regulated wage settlements, welfare state payments, government commitments to full employment and so on. These national systems of Fordism in turn are supported by an international system of regulation which guarantees free trade and fixed exchange rates, and which discourages extensive capital flows. Local systems of regulation take the form of Keynesianism, social democracy and the politics of the New Deal.

The Fordist system gave rise to a golden age of industrial capitalism in the 1950s and 1960s (Marglin and Schor 1990). Since that time, however, it has been unravelling under the weight of its internal contradictions. These contradictions include tendencies to declining rates of profit (as working-class power is institutionalized), to inflation and to excessive state intervention in the regulation of social life. To escape these contradictions two related and departing systems of regulation have evolved. On the one hand, the Fordist model is exported to some developing countries with the aim of establishing a system of global or peripheral Fordism. This process is based upon an extension of xenomonies which comes unstuck with the debt crises of the 1980s. In the heartlands of Fordism, meanwhile, there has been a movement in the direction of post-Fordism. Again, this system will vary from place to place and its minor contours are still not clear. At a minimum, post-Fordism would seem to involve the construction of privatized economies based around flexible production technologies, liberalization in international economic affairs (and a corresponding lack of coordination), and a new ethic of individualism based in consumerism. The boosters of post-Fordism might also link this new mode of regulation to the 'end of the Third World' and the 'end of history'.

The regulation school does not offer a fully formed theory of the dynamics of economic and social change under capitalism. Its models, indeed, are descriptive rather than explanatory. Nevertheless, in its theses we do find a tying together of certain threads we have met elsewhere. We are also provided with a template against which the more specific crises of debt and development can be located.

3   Radical Development Studies

The main propositions of radical political economy are linked very directly with a system-stability perspective on the debt crisis. For this reason, it makes sense to deal with these propositions at length and to say rather less about the related narratives of radical development studies. In any case, these narratives might be well known to the reader. Three prominent models within the radical tradition are (a) the development of underdevelopment model; (b) *dependencia* analyses; and (c) an account of the internationalization of capital and its effects on developing countries.

The Development of Underdevelopment

Radical theories of development and underdevelopment first took shape as an account of the development of underdevelopment. Paul Baran set the tone in his classic work on *The Political Economy of Growth*, first published in 1957 (Baran 1973). Baran reached back to Lenin (and to Dutt and Naoroji) to draw a distinction between a competitive capitalism which was more or less progressive, and a monopoly capitalism which is racked by over-accumulation (or by a lack of consumption in Baran's model). Monopoly capitalism is associated with a search for 'backward' countries and regions which can serve as an outlet for the goods and services of the metroplitan powers, and as a source of cheap labour and commodities. Baran argues that the so-called Third World is then created through a process of unequal exchange in an emerging capitalist world economy. Development and underdevelopment become two sides of the same coin in the nineteenth century, and the Third World is marked by a morphology of backwardness which is caused by a drain of resources from the periphery to the core of the world system.

This model was later generalized by the early work of Gunder Frank (and others). In a sense, Frank extended the zero-sum logics of Baran backwards in time, to the fifteenth, sixteenth, seventeenth and eighteenth centuries. Frank suggested that from 'the time of Cortez and Pizarro in Mexico and Peru, Clive in India, Rhodes in

Africa. . .the metropolis destroyed and/or totally transformed the earlier viable social and economic systems of these societies, incorporated them into the metropolitan dominated worldwide capitalist system, and converted them into sources for its own metropolitan capital accumulation and development' (Frank 1969, 225). These countries were exploited by means of systems of international exchange which were and are unequal in terms of labour-times transferred (Emmanuel, 1972), and which were and are policed by means of colonialism and neo-colonialism. This tradition of analysis, which includes the more sensitive analyses of Wallerstein, is often called neo-Marxist. This is because it adapts a Marxist account of exploitation at the point of production, centred in class relations, to fashion an account of exploitation in exchange between regions and countries. The most telling critique of this 'neo-Smithian Marxism' remains that of Robert Brenner (Brenner 1977).

Dependencia Analyses

The development of underdevelopment tradition of radical deve-lopment studies is sometimes counted as one part of a wider tradition of *dependencia* analyses. Gabriel Palma maintains that

> The general field of study of dependency analysis is the develop-ment of peripheral capitalism. Its most important contribution is its attempt to analyse it from the point of view of the interplay between internal and external structures. Its most well-known feature is the internal debate about whether capitalism remains 'historically progressive' in the Third World (i.e. capable of developing the productive forces of these societies and thus able to lead them towards socialism). (Palma 1989, 91; for more detail, see Palma 1978)

In addition to the development of underdevelopment narrative, Palma identifies three other *dependencia* analyses. A first narrative reaches back to Prebisch and reformulates a structuralist account of Latin American development. Like the development of underdeve-lopment tradition, this narrative is sceptical of the claim that the world system has so changed 'that the industrialization of the periphery [can] take place in the way predicted by Marx and

Engels' (Palma 1989, 91). This suggestion is in the manner of the duoeconomics identified by Hirschman (see chapter 3). The main point of focus of this narrative is on the terms of trade. Palma notes that the narrative was most actively propounded in the 1950s and 1960s, just before the terms of trade began to move in favour of Latin America. Cardoso remarked that, in this respect, 'history. . .prepared a trap for pessimists' (Cardoso 1977, 33).

Cardoso's remark also provides a pointer to rather different narratives within the *dependencia* tradition. His own work, in particular, presents dependency as an analysis of concrete processes of development (after Palma 1978). Cardoso is less minded than some of his colleagues to apply Marxism in a mechanical fashion, either to herald or to deride the possibility of capitalist industrialization in the periphery. His emphasis is on a partial and shifting dependency of particular regions and sectors in the periphery upon the capitalist world economy. His analysis is thus sensitive to the changing locational strategies of industrial and financial capital and to the locally determining effects of class formations and political alliances. His 'model' is altogether more iterative, more diffuse even, than are the models of the structuralists and the Frankians. His account is a possibilist account of dependency, development and underdevelopment (see Cardoso and Faletto 1979).

A fourth narrative within the *dependencia* tradition is focused on 'inhibited capitalist development in the periphery' (Palma 1989, 92). This analysis contains intimations of a later analysis of the articulation of modes of production (see Corbridge 1986), as does the work of Cardoso and his colleagues. The argument here is not that the periphery must remain forever as an outpost of underdeveloped *capitalism*, but that capitalism has not taken root, productively, in the peripheral economies. The development of capitalism has been blocked, according to this analysis, by an alliance of imperialist and pre-capitalist classes acting in tandem to inhibit the industrialization of peripheral economies. Development will come to these regions if, and only if, an incipient national bourgeoisie is able to assert itself and push ahead with local programmes of import-substitution industrialization.

The Internationalization of Capital

The *dependencia* school is sufficiently diverse that some of its claims overlap with a related analysis of the internationalization of capital. Briefly stated, the proponents of this view seek to reclaim Marxism from Lenin and to re-assert the progressive tendencies and possibilities inherent in capitalism (Warren 1980). According to this analysis, a binary division of the world economy into core and periphery is being displaced amidst a continuing process of time-space compression. As a consequence, the most pessimistic predictions of the development of underdevelopment school – that a capitalist industrialization of the periphery is logically impossible – have been disproved. Since the mid-1960s, industrial and financial capital have been exported in a manner that Baran and Frank could not have anticipated. Capital has embedded itself in new production spaces in order to escape the perils of working class militancy in the core and an associated decline in the rate of profit. As with capitalism generally, this process of establishment is exploitative: workers are exploited as the point of production, and nakedly so where surplus value is extracted in its absolute form. Where the internationalization of capital thesis breaks with some versions of dependency is in its suggestion that exploitation at the point of production *can* be 'developmental', in the sense of adding to growth and living standards. The presumption here is that it is better to be exploited by capital than not to be exploited at all (after Jenkins 1984, 51).

None of this means that capitalism is a perfect or even rational agency of development – far from it. The capitalist industrialization of the periphery is not only brutish in the short-term, it is also wildly unstable (Lipietz 1982). Capital is able to switch between production spaces with a degree of mobility which is unmatched by labour. The landscapes which are created in its image are thus quite uneven (Smith 1984; Brett 1985). They are also dependent upon flows of credit monies which are perilous in the extreme, both in terms of their availability and with respect to their need to be serviced. It is just this observation, of course, which brings us to the debt crisis.

4   Debt and Development Crises

From what has been said thus far, it will be apparent that a system-instability perspective on the debt crisis is less concerned with detailed, econometric relationships than it is with a broad political economy approach. The typical structure of a system-instability perspective on debt is as follows: (a) a critique of the standard narrative account of the debt crisis; (b) a discussion of the role of credit monies in the regulation of late capitalism; and (c) an analysis of the politics of monetary competition and crisis management.

Unmaking Myths

System-instability accounts of the 'debt crisis' are concerned to challenge a number of prevailing 'myths' about debt and development. A first myth has to do with the idea of *a* debt crisis. The singularity of this claim is attacked as misleading. To present the debt crisis as a precise event, or even as an interruption to normal economic life, is to give it a fixed location. *The* debt crisis then calls to mind Latin America in the 1980s, or the developing countries more widely. The possibility that *the* debt crisis is linked to the debt crises which affect the USA (external or internal) is correspondingly discounted, as are attempts to link *the* debt crisis to crises of credit money in the OECD more widely, or in Eastern Europe. The point of the critique, of course, is not to suggest that these crises are one and the same; it is simply to suggest that they are common crises which are being played out over an extended geographical stage.

A second myth has to do with the word 'debt'. Why a debt crisis, or why only a debt crisis? Why not a banking crisis or an export crisis? Why not a crisis of development, or of international money, or of Fordism? Again, the point is not a trite one. System-instability theorists are sensitive to the significance of naming and ordering. Naming is a first step to seeing. If we don't name the debt crisis as an export crisis, we might not see its unemployment implications in those regions not directly encumbered by debt. Similarly with the act of ordering: it is one thing to talk of debt and development crises, and quite another to talk of a debt crisis which has certain consequences for development. Words are weapons, as the old slogan has it; the use of words in debate and in discourse is not innocent and the act of naming must be attended to on that basis.

Finally, there is the matter of an origin for 'the debt crisis'. System-instability theorists are concerned that the debt crisis of the 1980s is traced back, very often, to the events of 1973 and 1974. In this manner, the debt crisis is represented as the result of an exogenous shock to an otherwise stable system (at least in the standard narrative account). Its origins lie also in the Third World, with OPEC. It is because of the actions of OPEC that petrodollar recycling becomes necessary and even possible. And it is because of OPEC, with the second oil shock, that sound-money policies are made necessary in the early 1980s, with all that this means for interest rates, recessions and debt servicing obligations.

The system-instability challenge to this view is not that it is wrong, in the sense of getting events in the wrong order, but that it is deceptive and thus politically corrupting. Robert Wood maintains that 'OPEC's role should not be overemphasized. . . .Euromarket expansion predated the OPEC price rises and in most years even after 1973 was based much more on non-OPEC than on OPEC deposits' (Wood 1986, 242). Branford and Kucinski agree, although their point of emphasis is elsewhere. They write of 'The myth of the oil-induced debt crisis', pointing out that 'only three of the ten most-indebted developing nations – Brazil, India and South Korea – depended heavily on imported oil' (Branford and Kucinski 1988, 64). In this manner they challenge the work of William Cline, when he argues that 'The single most important exogenous cause of the debt burden of non-oil developing countries is the sharp rise in the price of oil in 1973–74 and again in 1979–80' (Cline 1984, 8–9). Finally, there is a passage from a book authored primarily by Jackie Roddick which is worth quoting at length. Her statement speaks in support of the views of Wood and Branford and Kucinski, but it does so in a manner which speaks also to some of the main contours of a system-instability model of the developing countries' debt crisis. She writes:

> Economic historians usually blame the oil price rises of 1974 and 1979 for the problems facing Western industrial countries in the

1970s and 1980s. This is partly a myth, though a convenient one for politicians in the West. There were many reasons for the end of the long post-war boom, not least the breakdown in 1971 of the stable world trading system which had prevailed since Bretton Woods, due to the growing competition between the US and other powers. Rising inflation and declining levels of profit were caused not just by the OPEC price rises, but by the diversion of resources into international finance to cope with increasing levels of uncertainty in the international trading system. (Roddick 1988, 24)

Ponzi Schemes and Fictitious Capitals

A common point of focus for system-instability accounts of developing country debt are the contradictory relationships which exist (or can exist) between finance, accumulation and development. These relationships are explored in different ways. A populist, neo-Marxist line on the debt crisis explores the relationship of Ponzi schemes, so-called, to the local dynamics of mal-development in problem debtor countries. A more finely crafted account of this relationship explores the role of fictitious capitals in mediating the crises apparent in a prospectively global Fordist capitalism. More so than the populist account, this account pays some attention to class configurations in the indebted countries, and to the relationships between these configurations and local accumulation strategies.

Ponzi schemes and mal-development

A radical populist account of the debt crisis is set out in a number of texts, but most obviously in Susan George's book *A Fate Worse than Debt* (George 1989), and in Cheryl Payer's volume *Lent and Lost* (Payer 1991; see also Payer, 1974). Elements of this approach are apparent also in the work of Roddick (1988), Korner et al. (1986) and Branford and Kucinski (1988), although these texts are popularizing rather than populist. All five books are substantial contributions to the literature on the debt crisis.

The idea that the debt crisis is akin to a Ponzi scheme has been developed most explicitly by Cheryl Payer. Payer maintains that 'a "Ponzi scheme" denotes any scheme in which the original investors are paid off with money supplied by later investors. The same principle underlies "pyramid schemes" and chain letters. It is essential to our understanding of the Third World debt crisis to know that it, too, operated like those confidence schemes' (Payer 1991, 16).

By the standards of most writing on the debt crisis, this is a startling claim (though see Woo and Nasution 1989, 115; for a critique, see Díaz-Alejandro 1984). Payer develops the argument as follows: (a) development is defined in the post-war period as a process of economic growth based on capital-intensive industrialization; (b) to finance this development, resources external to a developing country are required to add to domestic savings; (c) most of these resources take the form of loans made to developing countries by official and private creditors. It is assumed that debts will be serviced on the basis of further loans being made to the indebted nation; (d) a model is then constructed which makes local processes of economic development dependent upon external sources of finance and purchases; (e) Payer denies that this model can be sustainable in the medium to long term. She uses the idea of a Ponzi scheme to argue that 'Either debt service would quickly rise to the point where it exceeded net lending or *new* lending must rise sharply in order to maintain a stable net transfer' (Payer 1991, 17: emphasis in the original). In either case the logic of 'development' is contradictory; (f) Payer concludes that the 'safety of the *lender's* capital was ultimately dependent. . .on *the willingness of new lenders to continue putting money into the country*' (ibid.; emphasis in the original). More importantly, the process of development is dependent on decisions which will be taken outside the developing country itself. If and when the plugs are pulled on the Ponzi scheme, it is left to the developing country to pick up the pieces. At this point the costs of (non-)development will be fought over by those agents which are party to the Ponzi scheme. The IMF, in particular, will police the debt crisis (so-called) in such a way that the burdens of devaluation will be borne by the indebted countries and by the poorest citizens within these countries. The well-to-do in the developing countries will usually act as *compradors* in this process; their main loyalty is to the transnational capitalist class to which they belong (see also Korner 1991).

It is important not to lose sight of the more general argument which emerges from Payer's account of the Ponzi scheme. Chapter 25 of Payer's book is entitled 'The myth of development through capital imports' and it is this aspect of her work which finds an echo in the work of Susan George. For George, like Payer, the debt crisis is symptomatic of a wider crisis of development, or what George calls the model of mal-development. Debts were accrued mainly because the West, and some local elites, were able to define the process of development in imitative terms. The mal-development model, says George, 'mimics without understanding and copies without controlling. Lacking roots in the local culture or environment, it quickly droops and withers if not sustained by transfusions – of foreign capital, technology and ideas. It goes for growth, usually without asking, "Growth of what? For whom?" Industrialization is frequently its centrepiece, sometimes export agriculture relying on industrial inputs' (George 1989, 14–15).

Put like this, of course, the work of George might not seem so very different to the work of Griffith-Jones and Sunkel, or even to that of some system-stability theorists who bemoan the urban–industrial bias of Keynesian development models. But this is not quite the case. George is against all forms of dependent development and she is insistent that spending on arms and ecocide has been central to most capitalist models of mal-development. Drawing on the work of the Stockholm International Peace Research Institute, George reckons that 'twenty per cent of Third World debt – OPEC excluded – can be attributed directly to arms purchases' (George 1989, 22). This point is not commonly made. Finally, George is scathing in her attack upon those creditors who are willing to lend to certain countries in Black Africa notwithstanding their appalling record on human rights. Although democratization and liberalization are the watchwords of the international creditor community, George notes that in practice loans will be made to a country like Zambia under Mobutu come what may; or at least as long as the country is not socialist. Indeed, the IMF is even willing to add insult to injury, demanding that the victims of indebted mal-development in Zaire are made to bear the further burdens of a process of structural adjustment.[4]

Fictitious capitals

The work of George and Payer is nothing if not committed. It is also radical-populist in the sense that the villains of the piece are clearly identified (the banks, the IMF and the USA), and insofar as the development of some regions and countries is predicated on the underdevelopment of some other regions and countries. Like most system-instability theorists, Payer and George are especially well informed on the profitability of the 'debt crisis' for the money-centre banks. In the case of Citibank, 72 per cent of overall earnings in 1976 came from international operations, with Brazilian operations alone accounting for 13 per cent of the bank's earnings (after Makin 1984, 133–4). Even 'Between 1982 and the end of 1985 profits at Banker's Trust went up by 66 per cent, at Chase Manhattan by 84 per cent, at Chemical by 61 per cent, at Citibank and Manny Hanny by 38 per cent and at Morgan Guaranty by 79 per cent' (George 1989, 39; see also table 5.1). George also refers to the banks' ability to avoid risk (in any meaningful sense), and to their capacity to engage successfully in the 'financial low intensity conflict [FLIC]' which she sees as a surrogate for open North/South warfare (see George 1989, 234–43; see also Raffer, 1989). She argues that FLIC is 'a process which allows the North to keep a check upon any pretensions to real independence on the part of the South and to ensure privileged access to the South's resources'

Table 5.1   Growth in foreign profits of US banks, 1970–1982

	Foreign profits (US$ million)			Foreign profits as % of total profits		
	1970	*1981*	*1982*	*1970*	*1981*	*1982*
Citibank	58	287	448	40	54	62
Bank of America	25	245	253	15	55	65
Chase Manhattan	31	247	215	22	60	70
Manufacturers Hanover	11	120	147	13	48	50
JP Morgan	26	234	283	25	67	72
Chemical New York	8	74	104	10	34	39
Bankers Trust New York	8	116	113	15	62	51

Source: After Roddick (1988, 33)

(George 1989, 234). FLIC is written through a conventional politics of debt management which ensures the profitability of the banks at the expense of the vulnerable in the South.

A second system-instability account of the debt crisis shares many of the conclusions of the mal-development theorists, but advances a more systematic account of the contradictions of capitalism in general (and of Fordism in particular). Marxists inclined to the views of the regulation scholars, or to Harvey's account of crises within capitalism, detect a sea-change in the organization of global capitalism in the late 1960s and 1970s. The debt crises are then symptomatic of this sea-change.

We can present this model in narrative form, the theoretical groundwork having already been laid. In the 1950s and 1960s global capitalism was organized around a long-standing division of labour which opposed a core to a periphery within a capitalist world economy. Apart from trade, the core maintained economic relations with the periphery through ODA and FDI. The international economic system was firmly policed by the USA, which was able to impose a clear set of rules as the hegemonic power. Some of these rules related to free trade and were delegated to the GATT and the IMF; still others had to do with the role of the dollar. The dollar, in the 1950s and 1960s, was more or less unchallenged as the *de facto* international unit of account. It was also the main source of international liquidity. At the same time, the dollar was as good as gold, in the sense that the Americans were bound not to undermine the value of the dollar. The Americans had agreed in 1944 to exchange dollars for gold at a fixed rate, with other currencies being pegged to the dollar. The Bretton Woods system in turn helped to maintain those national regimes of Fordism wherein mass production had been joined with mass consumption in a happy union. For the core Fordist powers, at least, this truly was the golden age of capitalism (Marglin and Schor 1990). By the mid-1960s it seemed that the business cycle had been eliminated by means of counter-cyclical policies of demand management.

We now know that this was not the case; the contradictions of capitalism were not buried in the 1950s. According to the account we are outlining, a crisis of over-accumulation became apparent in some of the primary circuits of capital from about the mid-1960s. In order to escape from these initial limits to capital, capital began to restructure, both within its core Fordist heartlands and by means of a new international division of labour. Capital now began to move offshore with a vengeance, both as production capital and as banking capital. The process was made possible by the emergence of new financial institutions and new circuits of monetary creation and transfer. Lipietz, in particular, makes the story of the xeno-markets central to his account of the (doomed) construction of a peripheral Fordism. He sees in Latin America and some NICs a spatial fix for the contradictions of central Fordism wherein future growth and development (and profits) would be prevalidated by the extension of private credit monies. In his view, the 'international debt economy was, then, based upon two assumptions. In terms of the creation of a mass of international money (xenocredits), it was assumed that capital investment in peripheral Fordism would prove profitable. In terms of the creation of an international money base (primary xenodollars), it was assumed that central Fordism would weather its balance of payments crisis' (Lipietz 1987, 143). The first of these assumptions was undone wherever capital was not put to productive use; wherever Pharaonism (that word again) was on the agenda, and where local regimes of accumulation were exclu-sionary and supportive of massive capital flight (see section 5.5 and table 5.2). Fictitious capitals in such cases were 'producing' fictitious spaces; either that, or they were producing built environ-ments in those favoured locations (Florida, New York, etc.) to which flight capital was attracted by way of the Caribbean's offshore banking centres (see Naylor 1987). The second assumption failed to materialize in the late 1970s and it was to address this crisis, in part, that 'central monetarism [was unleashed] to strangle peripheral Fordism' (Lipietz 1987, 160).

The OPEC price shocks are then a sub-plot in this story. In 1973/4 the USA responded to the first oil price rises by insisting that the way out of a looming crisis lay with a private recycling of the petrodollars transferred to OPEC. The Americans stood firm against an increase in SDRs at this time, or an extension in the old aid regime more generally. It was also supposed that banking and industrial capital would find more profitable investment opportuni-ties in the developing world than in an industrial world economy creaking under the weight of stagflation and strong trade unions (Peet 1987). The second oil-price shock of 1979–80 was significant

Table 5.2   Impact of capital flight on debt

	Gross external debt end-1985 (US$ billion)		Gross debt as percentage of exports of goods and services	
	Actual	Without capital flight	Actual	Without capital flight
Argentina	50	1	493	16
Brazil	106	92	358	322
Mexico	97	12	327	61
Venezuela	31	-12	190	-55
Malaysia	20	4	103	18
Nigeria	19	7	161	62
Philippines	27	15	327	195
South Africa	24	1	131	15

Source: Morgan Guaranty Trust Company, *World Financial Markets*, March 1986, table 12)

in lending strength to those forces in the industrial countries that wanted to end local systems of Fordism which infringed on the liberties enjoyed by capital. The 1979–80 crisis was contributory to the dawning of a new age of economic liberalism and privatization (or post-Fordism: see Lipietz 1987, chapter 6; and 1989).

David Harvey makes a similar set of points, but in different terms. Harvey focuses more directly upon the contradictory roles played by money and money capital in the processes of capital accumulation. He begins by noting that it is money which 'permits the separation of sales and purchases in space and time'. Money, like credit, is indispensable to the construction of a modern, capitalist society. But money can be issued and transferred in different forms and therein lies a contradiction. Briefly stated, money must fulfil two purposes. Money must first serve as a store of value and as a reliable indicator of relative prices. This function is most easily discharged by monies which are subjected to tight regulation, and/or which are supplied in a regular manner. The gold standard and the dollar-gold standard of the Bretton Woods era met these conditions. A second function of money is to finance the circulation of goods and services; in Marxist terms, 'By permitting fictitious capital to flourish, the credit system can support the transformation of circulating into fixed capital and meet the increasing pressures that arise as more and more of the total social capital in society begins to circulate in fixed form' (Harvey 1982, 269). This function of 'money' is best met by forms of paper money and private liquidity creation which are not subjected to close scrutiny. The monetary authorities are then faced with the task of trying to ensure that there is enough money in circulation to prevent a drift into economic depression, while making sure that there is not too much money in circulation that the value of money itself is debauched by inflation.

Harvey argues that this task cannot be resolved within capitalism; indeed, because 'the credit system is a product of capital's own endeavours to deal with the internal contradictions of capitalism' (Harvey 1982, 239; and consider the xenomarkets in particular), these contradictions are amplified and played out within the arena of money. The credit system only appears to buy off an underlying crisis of over-accumulation in the primary circuits of capitalism. In so doing, however, the crisis is generalized. This is evident in the case of the debt crisis or crises. Harvey suggests that the developing countries' debt crisis is one consequence of a process wherein 'the tendency towards excess in the realm of finance is ultimately checked by a return to the eternal verities of the monetary base' (Harvey 1982, 254). As such, the debt crisis in the Third World is linked to financial crises elsewhere, as in the USA where an internal debt of $11 trillion dwarfs an external debt itself equivalent to the total external debt of the developing world. Harvey joins with Lipietz in painting a picture of an unstable capitalism rebuilding itself in the 1970s on the basis of fictitious capitals extended to a Fordist periphery, only to collapse in the 1980s on the basis of bad debts in an era of tight money. At this point, though, the debt mountain moves elsewhere, as the USA, post-1982, is required to reflate its economy – again on the basis of debts acquired from abroad – to help ward off a generalized debt repudiation in Latin America. To ensure that US private capitals are not threatened by 'anti-capitalist' actions in Latin America, the USA itself becomes encumbered by debt as it seeks to maintain, or to renew, its role as global hegemon (Corbridge and Agnew 1991).

The analysis is simple but subtle at the same time. Capitalism is bound to live on tick, and yet parts of it must also die because of tick. To escape its internal contradictions, central Fordism is driven to purchase space and time (by credit monies), only to see the meagre 'development' which then ensues curtailed by a more elemental crisis in the function of money as a store of value. To put it one last way, the capitalist economy needs to become more like a casino economy (Strange 1986), but a casino economy itself must breed its corollary pathologies: the big swinging dicks, the dodgy banks (one thinks of BCCI), the uncertainties, the risks and the crashes. The Third World must live and die in the interstices of this 'system', as the creditors first extend credit monies and then demand repayment at high rates of interest. Some groups within the Third World are also party to the processes of local destruction which result. In the period since 1982 Mexico City has emerged as perhaps the largest financial services centre in Latin America. Mexican banks have played a major role in transferring Mexican and other Latin American funds abroad in the form of capital flight and allied hot money flows. Mexican banks have bought US banks in Texas and Florida, and have opened branches in the Cayman Islands and the Bahamas as well as on Wall Street.[5] As 'fixed points' in the modern capitalist system they have helped to speed up precisely that circulation of capital which see-saws in and out of Third World spaces at an alarming rate and to often debilitating effect.

Geopolitics

Thus far we have not mentioned politics or geopolitics. The model just outlined (which is a hybrid in any case) is an abstracted account of how and why the contradictions of capitalism are expressed in various monetary arenas. It puts the debt crisis into a very particular context. It also helps to explain when and why the Euromarket phenomenon first emerged and how it predates the activities of OPEC. But what of the historical geography of indebtedness? Why Latin America in particular? Why the US banks, and why a monetarist shock in the late 1970s courtesy of the USA? To answer these questions the system-instability perspective returns itself to that terrain where geopolitics intersects with political economy: the terrain of geopolitical economy. Again, we can present the main points in narrative form, paying particular attention to the roles played by the USA in a changing world economy.

In the 1950s the USA was unchallenged as the global hegemon, at least within the capitalist world. System-instability theorists are not alone in pointing to the great powers of the USA in relation to international economic, political, military and financial, affairs. The Bretton Woods system embodied rules that were mapped out by the Americans and the USA was prepared to police 'its' system as a benevolent despot (Parboni 1981, chapter 1). Parboni suggests that the USA had no need at this time to abuse the powers of financial seignorage which accrued to it as the country issuing the main international means of account. Indeed, the USA allowed other countries to pursue trading practices against it which were discriminatory and which impacted adversely on the US balance of payments. It did so because the main interest of the USA was in securing Western Europe and Japan (and parts of the Third World) as partners in an open international economy – and not as outposts of national capitalism (or, worse, socialism).

All this changed in the 1960s and 1970s. Parboni is not alone in arguing that the Bretton Woods settlement embodied a contradictory logic. The Bretton Woods system presupposed that the post-war world economy would develop evenly and by means of expanded trade flows between discrete national economies (Brett 1985). In a world of fixed ratios – rather than a shifting space of flows – the position of the USA would not then be challenged. The very success of the USA in rebuilding its allies, however (mainly by military expenditures justified in terms of the Cold War), meant that the Bretton Woods system could not continue in this form. In the 1960s the USA came to see its allies as economic rivals; it also saw capital drift from its shores in a new era of transnationalization and the export of capital. The USA was also faced with the growing costs attached to its role as global hegemon (Calleo 1987; Cox 1987). The war in Vietnam was proving very costly and the country faced a growing balance of payments crisis. The USA also faced a demand, from France in particular, that it honour its obligation to exchange gold for dollars. Gold drained from the USA in the 1950s and 1960s in a manner that only added to the gap which had been opening up between its reserves of gold and the number of dollars in circulation which were backed by this gold. (This is the so-called Triffin paradox: Triffin 1960.)

The system could not continue in this form. Parboni suggests that the Bretton Woods era was ended in August 1971 when President Nixon closed the gold–dollar window. This opened the way for the USA to print dollars to the extent that other countries were prepared to accept them as a means of payment. According to this interpretation, the 1970s were then set to become a decade of inflation mainly because the USA chose to restore its trading position by driving down the value of the dollar. Other countries were forced to respond to this strategy and a period of competitive devaluation ensued. This in turn made inflation worse and the process was not ended until confidence in the dollar was finally called into question in 1978. It was at this point that attempts were made to police a crisis of inflation by means of a return 'to the eternal verities of the monetary base', in the process transferring some of the costs of the crisis away from the countries that had conspired to create this mess in the first place (Tavares 1985; Tomassini 1984). The main instrument of this policing, as we have mentioned before, was the IMF – in Africa as in Latin America (see Onimode 1989a). The ideology behind which it took shelter was the ideology of the market.

In short, if there *is* an 'origin' for the developing countries' debt crisis it lies within the USA and with its pursuit of a politics of national exceptionalism in the 1970s. It was the USA that printed the dollars which could be on-lent to the Third World (or some parts of it); it was the USA that forced through a dramatic rise in the price of money in the early 1980s; and it was the USA that encouraged its private banks to recycle petrodollars following a first OPEC oil-price 'shock' (as opposed to agreeing to a programme 'of economic reflation or an increase in IMF and World Bank funding to counteract the effect of the OPEC price rises of 1974 and 1979': Roddick 1988, 19). (The USA was also ambivalent about the actions of OPEC: a rise in oil prices was likely to hurt Germany and Japan more than the USA; Parboni 1981, 52–4.) It then remained for the USA to police the debt crisis which broke out in Latin America in the 1980s. Now was the time, according to this account, for a rhetoric of belt-tightening to offset the excesses of the 1970s (notwithstanding that excess was a watchword for the USA itself in the 1980s). Now also was the time for victim-blaming, as a crisis-ridden system of unregulated capitalism was bought back to some point of temporary and cathartic 'equilibrium'. Just how this policing was put into effect (in the view of most system-instability theorists) is the subject of the next section. At this point, too, we outline an alternative prospectus for 'crisis management'.

5 Policing the Debt Crisis

System-instability theorists clear the decks for their debt management proposals by means of a critique of existing policies for policing the 'debt crisis'.

A first complaint is that the debt crisis has been policed by means of a policy of victim-blaming. The problem debtor countries are made responsible for the situation in which they find themselves; it is these countries and their citizens that must bear the brunt of structural adjustment programmes. System-instability theorists resist this logic. In their view, the mainsprings of indebtedness cannot derive from within the developing world. If any one country is required to shoulder the responsibility for twenty years of inflation, debts and deficits, it is the United States. The USA should be made to bear most of the costs of a crisis of US banking capital which has been policed as if it is mainly a developing countries' debt crisis.

This suggestion might seem to be at odds with the variable geography of indebtedness in the developing world, but this need not be the case. Lipietz draws an instructive contrast between inclusionary and exclusionary regimes of accumulation. He suggests that South Korea is exceptional because the Fordism which is established there is based on the true Fordist principle of an equation between domestic mass production and mass consumption. South Korea may be exclusionary in a political sense, but in economic terms it is inclusionary insofar as a majority of its population shares (unevenly) in the fruits of economic growth. This virtuous exchange in turn makes possible a system of development which is sustainable in economic terms and which allows outstanding debts to be serviced without great difficulty. South Korea is helped in this by its favourable integration into global trading circuits, and by the geopolitical support which it has consistently received from the USA.

The situation in Brazil and in most of Latin America is very different. Brazil exhibits the qualities of an exclusionary regime. Its Fordism was never secured domestically, but was imposed from outside in such a way that it would remain dependent and fragile as a result. Within the country vast inequalities remain and the majority of the population are unable to consume the goods and services produced in the Fordist sectors of the economy. Lipietz concludes that 'Because it was so exclusionary and therefore made the middle classes richer, the Brazilian regime generated a structural flow of imports by buying luxury goods or the means to produce them' (Lipietz 1987, 152). This is the class context within which we are to understand the political economy of capital flight. This is the context, too, for a wide-ranging account of the relationship between Pharaonism, mal-development and the servicing of external debt. Lipietz also notes that a similar situation might yet arise in India. Lipietz warns of 'the growing instability of Indian democracy; the centrifugal forces at work within it may well take on an ethnic or religious coloration, but they are based upon one of the most exclusive regimes of accumulation imaginable' (Lipietz 1987, 150). These remarks were written, one might suppose, in 1985–6.

By drawing attention to a possible future history of default in India, Lipietz is joining with other system-instability theorists in attacking a case-by-case approach to the policing of the debt crisis. Lipietz accepts that there are debt crises and debt crises, but he insists that the wider story *is* one of a global debt and development problem. System-instability theorists are keen to emphasize this point. The debt crises in Latin America and sub-Saharan Africa cannot be understood or managed effectively except in relation to the more profound disturbances which are afflicting a world economy in crisis. The case-by-case approach to debt management is misleading insofar as it fixes upon one symptom of this deeper malaise and not upon others (such as the US debt crises, or the Japanese trade surplus). It is also dishonest because a case-by-case approach to the debt crisis has been pursued more in word than in deed. Critics of a case-by-case approach maintain: (a) that the creditor community is well organized and has political and economic incentives to police the debt crisis by dividing to rule (see Korner et al. 1986, chapter 2); (b) that the rhetoric of case-by-case conceals a unity of approach whereby different developing countries are entreated to remarkably similar IMF structural adjustment programmes (the IMF, indeed, is derided for imposing misery and famine: Edwards 1988, 20); and (c) that resistance to a case-by-case approach is squashed by a mixture of carrot and stick policies, the carrot usually being dangled in front of Mexico to detach it from its erstwhile allies in a strategy of collective default (Fryer 1987), the stick being waved with sufficient crudeness that R. T. McNamar, US Deputy Treasury Secretary, can be quoted as saying: 'Have you ever contemplated what would happen to the president of a [repudiating] country if the government couldn't get insulin for its diabetics?' (Hector 1985, 27).

Finally, the containment-austerity years, and the Baker years, are said to conceal a hidden agenda to debt management which is rooted in US strategic concerns. This is most apparent in the desire of the USA to police a banking crisis as if it was solely a crisis of indebtedness, a point we have already made. Once the US banks were secure (in the mid-1980s) this policy could be relaxed and two others contemplated in its place. A first policy would be along the lines of the Bradley proposals and would respond to the needs of a particular fraction of US industrial capital. To the extent that funds for this proposal could be acquired from Japan, so much the better. The USA is not assumed to set the international economic agenda entirely on its own (especially when it is in debt to Japan and the rest of the world). A second policy would be to move in the direction of small-scale debt relief on the basis of conditionality (cf. the Brady proposals). The purpose of conditionality would then be to secure a progressive liberalization and privatization of the economies of the periphery, in the process making them attractive sites for US sales and investments. Debt–equity swaps would be the obvious instruments to secure this process of privatization, as they have been in Chile and Argentina. By the same token, organized labour movements would be among the first targets of this liberalization (Roxborough 1989).

There is clearly a good deal more that could be said about the system-instability critique of existing debt management strategies. We have yet to mention the work of Arthur MacEwan, for example, with regard to the absence of generalized debt repudiation in Latin America (MacEwan 1986). MacEwan contends that most Latin American countries have acquiesced to conventional debt management schemes because their local elites have a vested interest in preserving the status quo: their bank accounts are often in the USA and the bases of their political support are underpinned by the USA. (For a different view from within a system-instability perspective, see Branford and Kucinski (1988, 133–5). They maintain that the governments of most Latin American countries are Janus-faced, appearing pragmatic abroad and populist at home. In some cases the populist sentiment is minded more than in other cases – as in Garcia's Peru, or as in Brazil at the time of the Cruzado Plan. In such circumstances there is an opportunity for leftist forces within these countries to push at a door left half open by government. Thomas Hurtienne makes a similar point. He notes how a revamped ECLAC, and other Southern voices represented, for example, in the Cartegena Group, have argued recently for the repayment of 'legitimate debts' and the cancellation of 'non-legitimate' debts. Non-legitimate debts are debts incurred because of adverse changes in world economic conditions. In the case of Brazil, they amount to over one-half of its total external debt. The distinction allows populist governments in Latin America, and elsewhere, to face in two directions at the same time (Hurtienne 1991; see also Devlin, 1983, on what he calls bank 'monopoly rents').)

In general terms, however, the nature of the critique has been established. The system-instability perspective is critical of the rhetoric of 'no alternative' which surrounds the process of structural adjustment on a case-by-case basis. It calls attention to the political logics which inform this approach and to the economic pitfalls which must beset it as long as the more deep-rooted problems of dependent capitalism are left unattended (O'Connell 1985). Some system-instability theorists also write very movingly when they call to mind the circumstances which face the victims of the debt crisis. The work of Susan George continues to be notable in this respect. She reminds us that a decade of indebted underdevelopment has been associated with a large-scale destruction of the environment ('ecocide'), with growing hunger in parts of Africa and Latin America, and with swingeing cuts in some local health and education budgets. She also deals with these issues in the manner of an investigative journalist, substituting where necessary a biting ethnography for the anaemic prose of the 'dismal science' (economics). Whether the debt crisis is directly and wholly to blame for all of the evils which George describes is a moot point; none the less, George makes her readers think very carefully about the needs and rights of distant strangers and about the responsibilities of others to them (see also Onimode 1989b).

This brings us, finally, to some system-instability proposals for dealing with the debt crisis. It will be apparent that these can be more or less far reaching. The context for all of these proposals is a diagnosis of a sick and anarchic capitalism reeling from one crisis to another (Frank 1989). Some proponents of a system-instability perspective maintain that the debt crisis will not disappear as long as capitalism is: (a) unregulated; (b) based on the expansion and extension of fictitious capitals; and (c) subjected to geopolitical rivalries over the costs of devaluation (including those within the class-divided societies of the indebted Third World). More precisely, these theorists would call attention to the difficulties which face the indebted developing countries because of these instabilities, imbalances and rivalries. The US debts and deficits are one case in point. The USA now owes the rest of the world about $1000 billion. Interest payments are then of the order of $100 billion per annum. Although the USA can service these payments, the effects upon international trading patterns would be immense and depressive if the USA sought to finance its debts by means of a trade surplus. Such a surplus would be equivalent to 4–5 per cent of the value of total world trade. Nor is the prospect of the USA financing its debts by means of a general inflationism any more palatable, although this might entail a different set of winners and losers (Cohen 1989).

Wider issues, then, are not irrelevant to the more specific proposals which might be floated with regard to developing country debt (at least in system-instability terms). The whole thrust of the system-instability perspective is to draw attention to the deep-rooted instabilities which mark a global capitalism in transition. Nevertheless, some specific proposals have been floated by system-instability theorists and we can briefly outline three such proposals here.

1   A first 'proposal' comes from Lipietz. His suggestion is offered less by way of a cure than as a possible means out of the *current* crises of global Fordism. Lipietz suggests that, 'Given the present impasse, the most reasonable solution is to go on as usual, to pseudovalidate and to monetize debts in the hope that many of the most unwise investments will eventually find an outlet, and that the devalorization of the rest will be absorbed by a slight rise in world inflation' (Lipietz 1987, 185). Lipietz suggests that 'this could happen in several ways' (ibid.): (a) by a relaxation in the monetarist policies of the Federal Reserve; (b) by the setting up of a European credit pole alongside that of Japan – this would grant cheap or interest free loans to the developing world denominated in ecus; and (c) by the distribution of new credit monies (SDRs) to problem debtor countries free of charge, following a prior writing down of their bad debts.

2   A second set of proposals is embodied in Susan George's advocacy of a strategy for Debt, Development and Democracy. George says that her proposals amount to a call for a 'creative reimbursement' which would recognize that 'debt is not an economic but a political problem' (George 1989, 243). George once expressed a preference for a strategy of debt repudiation. She now suggests that this will not work unless all debtor countries agree on a total and collective repudiation of debts. The 3-D solution is thus a compromise. George continues:

> The basic idea of a 3-D solution is that countries are allowed to pay back interest and principal over a longer period of time in local currency, calculated so as not to create inflation. Their payments are credited to national development funds whose uses are determined by authentic representatives of the people working with those of the state. For the creditors, 3-D would come to the same thing as cancellation, since the local currency would be used internally. (George 1989, 244)

George insists that her proposals are not Utopian. The international financial system can now bear the costs of cancellation, and in the North 'environmentalists, peace activists, women's movements, trade unions, farmers and export-oriented industries. . .all have an interest in the changes that 3-D would encourage' (George 1989, 244).

3   A third set of proposals recommends confrontation and a concerted struggle against the evils of indebted underdevelopment. In effect, this is the 'Can't pay, Won't pay' strategy. It is a strategy which writes in support of women who have raided supermarkets in Colombia and elsewhere, helping themselves and their families to food on the basis of need (Walton, 1989). It is a strategy which commends the formation of activist alliances to press for debt repudiation and for an end to the ecocide and mal-development funded by debt and the negative net transfers to which it gives rise. It is a strategy which has been propounded by Castro (although not expounded by Cuba) and which found some sort of expression at the time of Garcia in Peru and amidst the 'IMF riots' which have affected countries as diverse as the Dominican Republic, Morocco and Jamaica. Repud:ation is commended as necessary and practicable; it is, indeed, put forward as a moral strategy, as a strategy which refuses to admit the legality of many of the debts contracted (often by military regimes), and which refuses to pay monetary debts in human lives.

6   Conclusion and Critique

The system-instability perspective on the debt crisis is distinguished by the broad temporal and spatial canvas on which it paints its accounts of debt and development. It is a perspective which begins its narratives in the 1960s and not in the 1970s, and which places the origins of the debt crisis in the North and not in the South. It is also a perspective on 'debt' which maintains that the developing countries' debt crisis is symptomatic of a deeper or wider malaise. Susan George continues this medical metaphor when she suggests that 'The debt crisis is a particularly ugly, acute manifestation of a chronic condition, the predictable outcome of economic strategies concerned far more with the world market than with local needs. Like an outbreak of carbuncles, it is spectacular on the surface but also a sure sign of underlying infection' (George 1989, 263).

The questions which might be asked of a system-instability perspective on the debt crisis vary between the populist and Marxist wings of this paradigm. A critique of the idea of a Ponzi scheme, for example – that it disregards the uses to which loans are put – will not hold in regard to the work of Harvey, Lipietz and others writing on the crisis tendencies inherent in Fordist capitalism. Nevertheless, there are some common questions which might be asked of a system-instability perspective.

First, what is the status of the concept of 'crisis' in the system-instability perspective, and what would a world look like that was not always in crisis? The point of this question is to render problematic the suggestion that crises might be absent from a more rationally organized society. If 'socialism' is the model for an alternative society, what status are we to grant those arguments which proclaim the death of socialism? Does an absence of crisis imply a steady state of sorts?

Second, and relatedly, if crises are the mainspring of economic change and 'development' in all modern societies, how and why should such crises be managed in ways which are different to those suggested by system-correction theorists? Isn't the system-instability perspective naive, or disingenuous, insofar as it attaches itself to debt management proposals which seek bluntly to write down debt in favour of development? Where does this leave the notion of responsibility? How does it speak to the left's own recognition that the governments of many indebted countries are far from benign and are likely to be maintained in power by conciliatory defaults and/or pre-emptive concessional debt write-downs? How can this approach be defended against a system-stability perspective which maintains that hard choices have to be made: that meeting local needs in the short-term might jeopardize a faster and more sustainable pattern of development in the long-term? Is the 'Third Worldism' of some sections of the (populist) left not at odds with (or uniformed by) more contemporary debates on the question of sovereignty in the developing world?

Third, what is the precise nature of the development 'model' which is meant to substitute for the pernicious mal-development of the past? Is this to be a form of planned development in which a visible hand substitutes for an invisible hand – and if so, whose visible hand and on what authority? Is development best served by a politics of localism which seems to turn its back on the pleasures as well as the pains of a market-led modernization? Isn't this an invitation to a lack of transparency in local economic and political affairs?

The list of questions can be extended, but some idea of the 'case to answer' will be apparent. It should also be evident from this chapter that answers will be forthcoming, and forthcoming in terms that the reader should be able to anticipate (in part) and judge for himself or herself. This has been the main theme of the book. There is a close link between policies which are proposed to deal with the 'debt crisis' and the discourses on debt and development which make these policies possible. The conclusion to the book dwells further on this theme while rendering certain aspects of it problematic and open to argument.

Notes

1   This chapter cannot hope to do justice to several new varieties of neo-Marxism and post-Marxism. Suffice it to say that methodological individualism is not unknown within Marxism: see Carling (1986) on 'rational choice Marxism'; see also the work of John Roemer (1981, 1988) and Jon Elster (1985). For a critique, see Wood (1989).

2   In absolute terms, the model is not a zero-sum model. It is clearly possible for both capitalists and workers to increase their profits and wages in absolute terms, even if one class is able to increase its rate of return at the expense of the other.

3   This may suggest an endemic underconsumption within capitalism. A more sophisticated account of the accumulation process under the rule of capital would want to pay attention to the ways in which capital is switched between various departments of production (see Howard and King 1975) and between the various circuits of capital (see the next sub-section).

4   This criticism may be insensitive to a recent debate within the IMF and the World Bank about structural adjustment lending, conditionality and local processes of democratization. At the centre of this debate is the concept of sovereignty: under what circumstances is it legitimate for an institution such as the World Bank to seek to make funds available to a country only when that country first undertakes to reform its record with respect to abuses of civil and human rights (for example)? The debate is not an easy one and Southern voices are themselves divided about this new twist in a long-standing debate on conditionality. For the wider debate about democratization, see Lehmann (1990).

5   I am indebted to Dr Gareth Jones of the University College of Wales, Swansea, for the information on Mexico City.

SIX

Conclusion

Debt and Development has not tried to present a detailed empirical investigation of the dynamics of indebtedness in different regions and countries of the developing world (for examples of such work, see Kraft 1984; Felix 1987; Frieden 1987b; Onimode 1989a). Nor has this book sought to contribute to an important and continuing debate on the formulation of debt management policies (see Bird 1987; Griffith-Jones 1988, 1991); it is not a practical book in this sense. *Debt and Development* has rather aimed to provide a narrative account of the debt crisis, together with three particular readings of the nexus of issues which define the relationships between debt, development and mal-development. In the light of these aims, two questions remain to be discussed. First, what, if any, are the relationships which exist between particular readings of the debt crisis and particular strategies of debt crisis management? Second, how, if at all, can we adjudicate between the competing claims and narratives advanced in chapters 3–5?

The question of the relationship between narratives and debt management practices returns us to Hirschman and his warning that 'it is difficult – and often ludicrous – to assign intellectual responsibility for actual policy decisions, let alone policy outcomes' (Hirschman 1981, 111). But it also returns us to Keynes and to his suggestion that 'Practical men, who believe themselves to be quite exempt from any intellectual influences, are usually the slaves of some defunct economist. Madmen in authority, who hear voices in the air, are distilling their frenzy from some academic scribbler of a few years back' (Keynes 1936, 383).

If this sounds cryptic, let me rephrase it. One conclusion which should not be drawn from *Debt and Development* is that the developing countries' debt crisis has been managed entirely in accordance with the claims and assumptions set out by system-stability theorists and/or by system-correction theorists. It is rather the case that debt management practices have shifted in the 1980s, partly in response to previous policies and practices, partly in recognition of the changing position of the commercial banks, and partly in deference to changing geopolitical realities. Put crudely, we might say that between 1982 and 1985 the debt crisis was managed as a banking crisis. At this time, too, the USA took a lead in shaping debt management policies – moving to protect its commercial banks and to strike deals which would favour its neighbour Mexico in relation to more distant debtor countries. In the mid-1980s this attitude began to shift. As the US economy moved out of recession, so the Baker plan promised to couple debt crisis measures to a new age of growth and export expansion in the world economy. More significantly, the commercial banks began to rebuild their capital bases. In the wake of actions taken by Peru and Brazil between 1985 and 1987, the banks also moved to make provisions for possible bad debts. This non-governmental approach to the debt crisis at last began to undo the stand-off which had been apparent since 1982. New instruments for debt conversion and debt write-downs now appeared, and the Brady plan was put into place to add some sort of official imprimateur to a process of debt management already underway by dint of private initiatives. By 1990, approaches to the debt crisis had become a good deal more flexible, with the Japanese, the French and the British all taking actions, and/or pushing the USA to take actions, which would reduce the outstanding stocks of private and official debt in Latin America and sub-Saharan Africa.

This description of events does not sit easily with the narratives set out in chapters 3, 4 and 5 of this book. But this does not mean that these narratives are of academic interest alone, or that they are at odds with the realities of contemporary debt management practices. A more reasonable conclusion to draw is that elements of the system-stability and system-correction narratives have been deployed, in different contexts and at different times, *as part of* the debt management practices of the 1980s. System-instability ideas have also been raised as part of the negotiating stance of some debtor countries.

In a sense this takes us back to the work of Giddens and the structuration theorists; to the suggestion that economic and political ideas are always vital to – *and part of* – the production and reproduction of systems of economic organization and management. Expert systems are part and parcel of the process of social production. More directly, we can suggest that the containment-adjustment years of 1982–5 were strongly informed by system-stability ideas (especially at the level of public relations and rhetoric), even as they were informed also by the geopolitical agendas of the USA and by the practical needs of the commercial banks. The idea that the debt crisis was a temporary and localized interruption to the workings of an otherwise sound world economy was promoted with vigour at this time by the World Bank, just as the discourse of profligacy featured strongly in the debt crisis negotiations of the mid-1980s. The concept of moral hazard is another which was advanced to telling effect by the creditors, together with the suggestion that there was no alternative to the rigours of structural adjustment.

In the second half of the 1980s debt management practices drew more eclectically from the several discourses advanced by system-stability and system-correction theorists. At this point the global consequences of a supposedly local crisis of indebtedness were registered more directly in the official discourses of the creditor powers, and a new recognition emerged that the debt crisis would not be resolved in the absence of concerted international action to provide for global economic growth. The voices of business groups were also more active at this time, as were those voices calling for default in the indebted world. This changing political landscape made it difficult for some creditor organizations to persist with a discourse of debt management that equated the debt crisis more or less exclusively with a banking crisis (Milivojevic 1985). A more diverse set of management policies was now bound to emerge, and would draw upon accounts of the debt crisis which were less certain with regard to its causes, consequences and supposed moral verities. Nevertheless, a strong dose of economic orthodoxy remained the order of the day, particularly in regard to structural adjustment programmes and the need for economic liberalization.

If it is unwise to read off particular policies from particular theories of the debt crisis, it is just as inadvisable to underestimate the power of the counter-revolution in development theory and policy in the 1980s. The counter-revolution was especially successful in defining the limits of what was – and what was not – possible and desirable in the realm of economic and political activities in the developing world.

But what *is* desirable, and how are we to adjudicate between the competing claims and strategies put forward by economic theorists and by economic and political managers? What can an analysis of the different discourses on debt and development contribute to these intellectual and policy-related debates? These questions are not unrelated to our first set of questions, but their point of focus is different. Our concern now is less with the links which exist between theory and policy, than with the academic task of seeking a reasonable explanation of the world and of particular events and processes therein. Put bluntly, the question now is: 'which of chapters 3, 4 and 5 offers the most convincing account of the crises of debt and development?'

The answer, of course, is that the question is badly put. Epistemological and ideological purists will dislike this suggestion, and the charge of eclecticism may well be raised in some quarters. I remain convinced, however, that each of the narrative accounts set out in chapters 3, 4 and 5 has its merits, that these three sets of accounts are not mutually exclusive of each other's claims in all respects (notwithstanding some obvious and important differences), and that the question of 'good or bad' is meaningful only with regard to some more specific set of criteria.[1] In part, this has to do with the temporal and spatial parameters of the three sets of perspectives themselves, a point I have tried to bring out at the end of each of chapters 3–5. Thus, some radical accounts of the debt crisis are attractive insofar as they are willing to fashion an explanation of the debt crisis which ranges over a broad spatial and temporal canvass. The radical account is effective in the way in which it calls into question particular strategies of development/mal-development, and in the manner in which it is able to map out certain instabilities and crisis tendencies in the process of capital accumulation itself. Narrow the horizons, though, and the practical suggestions and implications of this paradigm are less easy to fathom. Precisely because systemic change is called for, it is not clear what second-best policies might be enacted in a manner which is consistent with the system-instability perspective. It is not clear either that an account of capitalism in perpetual crisis – this being its natural state – is best suited to a practical negotiation of the particularities of the debt crisis. (This applies less to the work of some Marxists than it does to the work of the radical populists).

The system-stability account is weak and strong for much the same reasons. Its strength is its vision: its normative account of an efficient and free world market delivering employment, choice and higher incomes to most of those participating in the global economy. Its weakness is its unwillingness to deal with the contradictions of an imperfect world – of a world of economics 'tainted' by a world of politics. Too often, its suggestions are likely only to be punitive, at least in the short to medium term. Once again, the end can come to justify the means (perversely so, given a disposition in this quarter to despise end-state arguments), with the poor and the powerless being made to bear the costs of 'perfect policies' in an imperfect, uneven and divided world political economy. Precisely because people are not able to enter the markets as equals, it is difficult to see how or why *the market* should be looked upon as their guardian and saviour.

This leaves us with the system-correction perspective. Persuaded as I am by the Marxissant account of the contradictions of capitalism and modernity, these same contradictions and trade-offs persuade me that proposals for economic reform and management must acknowledge these tensions and work within them. This need not be a defence of the status quo, for by working within – or with regard to – these constraints, the 'system' is slowly but surely changed (it changes anyway). It is, however, an argument against intellectual and political finalism. It is an argument against accounts of the world economy which assume that the contradictions of capitalism and modernity can be neatly transcended – by free markets and/or by socialism – as opposed to modified and replaced by others which might be less pernicious and less asymmetrical.

This, it seems to me, is the place to end this book: not with a detailed prospectus for debt management reforms (which would be out of place), but with the suggestion that development is defined by its contradictions and by its dilemmas (to return to John Toye's helpful phrase). To expect simple answers to complex questions is facile and misleading. Uncertainty is the nature of the modern condition and a celebration of this fact at the level of theory and policy might be no bad thing. But uncertainty is not to be confused with intellectual or moral relativism. My argument is not an argument for inaction, or for policies which fail to register the evident injustices of the modern world system. It is, rather, an argument in favour of discrimination; an injunction always to be critical of particular stories about debt and development, and always to be cautious in proposing solutions to problems which resist simple diagnoses. It is for this reason that I am drawn to system-correction accounts of the debt crisis, even though I do not discount several of the insights and questions raised by proponents of system-stability and system-instability perspectives on debt and development. The need to choose sharply between these three points of view does not always arise, notwithstanding their different models of the 'good society'. The willingness of some authors to construct narratives which claim to be opposed in most respects to the narratives of others does not imply that the person reading these texts is obliged to make a similar choice. The certainty of choice does not mean that we have to choose certainty; uncertainty, or an attitude of radical doubt, is not something to be despised.

Note

1 I do not want this suggestion to be misunderstood. I accept, of course, that the grand narratives set out in chapters 3 and 5 (especially) are incommensurable to the extent that each is constructed on the basis of a singular conception of the good society. That said, I do not accept that all or even most of the claims developed in support of these grand visions are also incommensurable. Certain propositions are common to all discourses, to varying degrees, and can be interrogated in terms of certain common discursive properties (for example, of empirical and logical consistency). Thus, while system-stability and system-instability views on the significance and extent of capital flight clearly differ, they do not depend fully, or even in the main, upon the ontological propositions of each of these perspectives.

Bibliography

Abbott, G. (1989) Another round of retroactive terms adjustment. In H. O'Neill (ed.), *Third World Debt: How Sustainable Are Current Strategies and Solutions?* London: Frank Cass, 60–77.

Amsden, A. (1989) *Asia's Next Giant.* Oxford: Oxford University Press.

Anayiotis, G. and de Piniés, J. (1990) The secondary market and the international debt problem. *World Development*, 18, 1655–69.

Bacha, E. (1987) IMF conditionality: conceptual problems and policy alternatives. *World Development*, 15, 1457–67.

Bailey, N. (1983) A safety net for foreign lending. *Business Week*, 10 January.

Baker, J. (1985) Statement before the Joint Annual Meeting of the IMF and the World Bank, Seoul, South Korea, 8 October.

Balassa, B. (1981) *The Newly Industrialising Countries in the World Economy*, Oxford: Pergamon.

Baran, P. (1973) *The Political Economy of Growth*, Harmondsworth: Penguin.

Bauer, P. (1972) *Dissent on Development.* London: Weidenfeld and Nicolson.

Bauer, P. (1981) *Equality, the Third World and Economic Delusion.* London: Methuen.

Bauer, P. (1984) *Reality and Rhetoric: Studies in the Economics of Development.* London: Weidenfeld and Nicolson.

Bauer, P. (1991) *The Development Frontier: Essays in Applied Economics.* Hemel Hempstead: Harvester Wheatsheaf.

Beenstock, M. (1984) *The World Economy in Transition.* London: George Allen and Unwin.

Bird, G. (1987) Interest rate compensation and debt: would a cap fit? *World Development*, 15, 1237–42.

Bradley, B. (1986) A proposal for Third World debt. Speech delivered in Zurich. 29 June.

Brandt, W. (Chairman of the Brandt Commission) (1983) *Common Crisis, North-South: Cooperation for World Recovery*. London: Pan.

Branford, S. and Kucinski, B. (1988) *The Debt Squads: the US, the Banks and Latin America*. London: Zed.

Brenner, R. (1977) The origins of capitalist development: a critique of neo-Smithian Marxism. *New Left Review*, 104, 25–92.

Brett, E. A. (1985) *The World Economy since the War: the Politics of Uneven Development*. London: Macmillan.

Buiter, W. and Srinivasan, T. (1987) Rewarding the profligate and punishing the prudent and poor: some recent proposals for debt relief. *World Development*, 15, 411–17.

Buiter, W., Kletzer, K. and Srinivasan, T. (1989) Some thoughts on the Brady plan: putting a fourth leg on the donkey? *World Development*, 17, 1661–4.

Calleo, D. (1987) *Beyond American Hegemony: the Future of the Western Alliance*. New York: Basic Books.

Cardoso, E. and Dornbusch, R. (1989) Brazilian debt crises: past and present. In B. Eichengreen and P. Lindert (eds), *The International Debt Crisis in Historical Perspective*. Cambridge, MA: MIT Press, 106–39.

Cardoso, E. and Fishlow, A. (1989) The macroeconomics of the Brazilian external debt. In J. Sachs (ed.), *Developing Country Debt and the World Economy*. Chicago: University of Chicago Press/NBER, 81–99.

Cardoso, F. H. (1977) The consumption of dependency theory in the US. *Latin American Research Review*, XII, 7–24.

Cardoso, F. H. and Faletto, E. (1979) *Dependency and Development in Latin America*. Berkeley: University of California Press.

Carling, A. (1986) Rational choice Marxism. *New Left Review*, 160, 24–62.

Carmichael, J. (1989) The debt crisis: where do we stand after seven years? *World Bank Research Observer*, 4, 121–142.

Chenery, H., Robinson, S. and Syrquin, M. (1986) *Industrialization and Growth: a Comparative Study*. Oxford: Oxford University Press/World Bank.

Chenery, H. and Syrquin, M. (1975) *Patterns of Development, 1950–1970*. London: Oxford University Press.

Claessens, S. and Diwan, I. (1990) Investment incentives: new money, debt relief and the critical role of conditionality in the debt crisis. *World Bank Economic Review*, 4, 21–41.

Clausen, W. (1983) Let's not panic about Third World debts. *Harvard Business Review*, 61, 106–14.

Cline, W. (1984) *International Debt: Systemic Risk and Policy Response.* Washington, DC: Institute for International Economics.

Cohen, J. (1989) The US as world's No. 1 debtor: causes and consequences. In H. Singer and S. Sharma (eds), *Economic Development and World Debt.* London: Macmillan, 355–67.

Cole, K., Cameron, J. and Edwards, C. (1983) *Why Economists Disagree: the Political Economy of Economics.* Harlow: Longman.

Cole, S. (1989) World Bank forecasts and planning in the Third World. *Environment and Planning A*, 21, 175–96.

Congdon, T. (1988) *The Debt Threat.* Oxford: Blackwell.

Corbo, V. and de Melo, J. (eds) (1985) Special issue on 'Liberalization and stabilization in the Southern Cone of Latin America'. *World Development*, 13 (8).

Corbridge, S. (1986) *Capitalist World Development: a Critique of Radical Development Geography.* London: Macmillan.

Corbridge, S. (1988) The debt crisis and the crisis of global regulation. *Geoforum*, 19, 109–30.

Corbridge, S. and Agnew, J. (1991) The US trade and budget deficits in global perspective: an essay in geopolitical-economy. *Society and Space*, 9, 71–90.

Corden, M. (1988) Debt relief and adjustment incentives. *International Monetary Fund Staff Papers*, 35, 628–43.

Corden, M. (1991) The theory of debt relief: sorting out some issues. *Journal of Development Studies*, 27, 133–45.

Cornia, G., Jolly, R. and Stewart, F. (eds) (1987) *Adjustment with a Human Face: Protecting the Vulnerable and Promoting Growth.* Oxford: Oxford University Press.

Cox, R. W. (1987) *Production, Power and World Order: Social Forces in the Making of History.* New York: Columbia University Press.

Dale, R. (1985) *The Regulation of International Banking*, Cambridge, MA: Woodhead-Faulkner.

Daly, M. and Logan, M. (1989) *The Brittle Rim: Finance, Business and the Pacific Region.* Harmondsworth: Penguin.

Devlin, R. (1983) Renegotiation of Latin America's debt. *Cepal Review*, 20, 101–12.

Díaz-Alejandro, C. (1983) Stories of the 1930s for the 1980s. In P. Aspe, R. Dornbusch and M. Obstfeld (eds), *Financial Policies and the World Capital Market: the Problem of Latin American Countries.* Chicago: University of Chicago Press, 5–40.

Díaz-Alejandro, C. (1984) Latin American debt: I don't think we are in Kansas anymore. *Brookings Papers on Economic Activity*, 2, 335–89.

Dornbusch, R. (1985) Policy and performance links between LDC debtors and industrial countries. *Brookings Papers on Economic Activity*, 2, 303–56.

Dornbusch, R. (1986) *Dollars, Debts and Deficits*. Cambridge, MA: MIT Press.

Economic Commission for Latin America and the Caribbean (1989) *Economic Survey of Latin America and the Caribbean, 1988*. Santiago: United Nations.

Economist (1982) The crash of 198? 16 October.

Economist (1989) A survey of the Third World. 23 September.

Edwards, C. (1985) *The Fragmented World: Competing Perspectives on Trade, Money and Crisis*. London: Methuen.

Edwards, C. (1988) The debt crisis and development: a comparison of major competing theories. *Geoforum*, 19, 3–28.

Edwards, S. (1989) Structural adjustment policies in highly indebted countries. In J. Sachs (ed.), *Developing Country Debt and the World Economy*. Chicago: University of Chicago Press/NBER, 249–62.

Eichengreen, B. and Lindert, P. (eds) (1989a) *The International Debt Crisis in Historical Perspective*. Cambridge, MA: MIT Press.

Eichengreen, B. and Lindert, P. (1989b) Overview. In B. Eichengreen and P. Lindert (eds), *The International Debt Crisis in Historical Perspective*. Cambridge, MA: MIT Press, 1–11.

Eichengreen, B. and Portes, R. (1989) After the deluge: default, negotiation, readjustment. In B. Eichengreen and P. Lindert (eds), *The International Debt Crisis in Historical Perspective*. Cambridge, MA: MIT Press, 12–47.

Elster, J. (1985) *Making Sense of Marx* Cambridge: Cambridge University Press.

Emmanuel, A. (1972) *Unequal Exchange*. London: Monthly Review Press.

Fei, J., Ranis, G. and Kuo, S. W. (1980) *Growth with Equity: the Taiwan Case*. Oxford: Oxford University Press.

Feldstein, M. (1987) Latin America's debt: muddling through can be just fine. *Economist*, 27 June.

Felix, D. (1987) Alternative outcomes of the Latin American debt crisis: lessons from the past. *Latin American Research Review*, 22 (2), 3–46.

Ffrench-Davis, R. (1990) Debt–equity swaps in Chile. *Cambridge Journal of Economics*, 14, 109–26.

Findlay, A. (ed.) (1988) *Debt Stabilization and Development*. Oxford: Blackwell.

Fishlow, A. (1985) Lessons from the past: capital markets during the nineteenth century and the interwar period. *International Organization*, 39, 383–439.

Frank, A. G. (1969) *Latin America: Underdevelopment or Revolution.* London: Monthly Review Press.

Frank, A. G. (1989) Debt where credit is due. In H. Singer and S. Sharma (eds), *Economic Development and World Debt.* London: Macmillan, 33–8.

Frieden, J. (1987a) *Banking on the World: the Politics of International Finance.* New York: Harper and Row.

Frieden, J. (1987b) The Brazilian borrowing experience: from miracle to debacle and back. *Latin American Research Review,* 22 (1), 95–131.

Friedman, M. (1960) *A Program for Monetary Stability.* New York: Fordham University Press.

Friedman, M. (1969) *The Optimum Quantity of Money and Other Essays.* Chicago: Aldine.

Friedman, M. and Friedman, R. (1980) *Free to Choose.* Harmondsworth: Penguin.

Fryer, D. (1987) The political geography of international lending by private banks. *Transactions of the Institute of British Geographers,* 12, 413–32.

Fukuyama, F. (1989) The end of history? *The National Interest,* Summer, 3–18.

Furtado, C. (1963) *The Economic Growth of Brazil.* Berkeley: University of California Press.

Gamble, A. (1988) *The Free Economy and the Strong State: the Politics of Thatcherism.* London: Macmillan.

Garrison, R. and Kirzner, I. (1989) Friedrich August von Hayek. In J. Eatwell, M. Milgate and P. Newman (eds), *The New Palgrave: the Invisible Hand.* London: Macmillan, 119–30.

George, S. (1989) *A Fate Worse than Debt.* Harmondsworth: Penguin.

Green, R. (1989) The broken pot: the social fabric, economic disaster and adjustment in Africa. In B. Onimode (ed.), *The IMF, the World Bank and the African Debt: Volume 2 – the Social and Political Impact.* London: Zed, 31–55.

Greider, W. (1987) *Secrets of the Temple: How the Federal Reserve Runs the Country.* New York: Simon and Schuster.

Griffith-Jones, S. (ed.) (1988) *Managing World Debt.* Hemel Hempstead: Harvester Wheatsheaf.

Griffith-Jones, S. (1991) Creditor countries' banking and fiscal regulations: can changes encourage debt relief? *Journal of Development Studies,* 27, 167–91.

Griffith-Jones, S. and Sunkel, O. (1986) *Debt and Development Crises in Latin America: the End of an Illusion.* Oxford: Clarendon Press.

Grimwade, N. (1989) *International Trade: New Patterns of Trade, Produc-*

tion and Investment. London: Routledge.

Guttentag, J. and Herring, R. (1985) *The Current Crisis in International Banking*. Washington, DC: Brookings Institution.

Gwynne, S. (1983) Adventures in the loan trade. *Harper's*, September.

Haberler, G. (1959) *International Trade and Economic Development*. Cairo: National Bank of Egypt.

Haberler, G. (1985) *The Problem of Stagflation: Reflections on the Microfoundation of Macroeconomic Theory and Policy*. Washington, DC: American Enterprise Institute.

Haberler, G. (1987) Liberal and illiberal development policy. In G. Meier (ed.), *Pioneers in Development (Second Series)*. Oxford: Oxford University Press/World Bank, 51–83.

Haggard, S. (1986) The politics of adjustment: lessons from the IMF's Extended Fund Facility. In M. Kahler (ed.) *The Politics of International Debt*. Ithaca, NY: Cornell University Press, 157–86.

Haggard, S. and Kaufman, R. (1989) The politics of stabilization and structural adjustment. In J. Sachs (ed.), *Developing Country Debt and the World Economy*. Chicago: University of Chicago Press/NBER, 263–74.

Hall, P. (ed.) (1989) *The Political Power of Economic Ideas: Keynesianism across Nations*. Princeton, NJ: Princeton University Press.

Harrod, R. (1972) *The Life of John Maynard Keynes*. Harmondsworth: Pelican.

Harvey, D. (1982) *The Limits to Capital*. Oxford: Blackwell.

Harvey, D. (1989) *The Urban Experience*. Oxford: Blackwell.

Hayek, F. A. (1944) *The Road to Economic Serfdom*. Chicago: University of Chicago Press.

Hayek, F. A. (1960) *The Constitution of Liberty*. London: Routledge and Kegan Paul.

Hayek, F. A. (1976) *Denationalization of Money*. London: Institute of Economic Affairs.

Hector, G. (1985) Third World debt: the bomb is defused. *Fortune*, 18 February, 24–9.

Hill, P. (1986) *Development Economics on Trial: the Anthropological Case for a Prosecution*. Cambridge: Cambridge University Press.

Hirschman, A. (1981) *Essays in Trespassing: Economics to Politics and Beyond*. Cambridge: Cambridge University Press.

Howard, M. and King, J. (1975) *The Political Economy of Marx*. Harlow: Longman.

Hunt, D. (1989) *Economic Theories of Development*. Hemel Hempstead: Harvester Wheatsheaf.

Hurtienne, T. (1991) Is there a way out of the crisis for the indebted capitalist world? In E. Alvater, K. Hubner, J. Lorentzen and R. Rojas (eds), *The Poverty of Nations: a Guide to the Debt Crisis from Argentina to Zaire*. London: Zed, 105–24.

IMF (1988) *International Financial Statistics*. Washington, DC: International Monetary Fund.

Jenkins, R. (1984) Divisions over the international division of labour. *Capital and Class*, 22, 28–57.

Jessop, B. (1982) *The Capitalist State*. Oxford: Martin Robertson.

Johnston, B. and Kilby, P. (1975) *Agriculture and Structural Transformation*. Oxford: Oxford University Press.

Kaletsky, A. (1985) *The Costs of Default*. New York: Twentieth Century Fund.

Kaufman, R. (1990) Stabilization and adjustment in Argentina, Brazil and Mexico. In J. Nelson (ed.), *Economic Crisis and Policy Choice: the Politics of Adjustment in the Third World*. Princeton, NJ: Princeton University Press, 63–111.

Kenen, P. (1983) A bailout plan for the banks. *The New York Times*, 6 March.

Keynes, J. M. (1936) *The General Theory of Employment, Interest and Money*. London: Macmillan.

Keynes, J. M. (1971) *The Economic Consequences of the Peace*. London: Macmillan.

Killick, T. (ed.) (1984) *The Quest for Economic Stabilisation: the IMF and the Third World*. London: Heinemann/ODI.

Kirzner, I. (1989) *Discovery, Capitalism, and Distributive Justice*. Oxford: Blackwell.

Korner, P. (1991) Zaire: indebtedness and kleptocracy. In E. Altvater, K. Hubner, J. Lorentzen and R. Rojas (eds), *The Poverty of Nations: a Guide to the Debt Crisis from Argentina to Zaire*. London: Zed.

Korner, P., Maass, G., Siebold, T. and Tetzlaff, R. (1986) *The IMF and the Third World*. London: Zed.

Kraft, J. (1984) *The Mexican Rescue*. New York: Group of Thirty.

Krueger, A. (1974) The political economy of the rent-seeking society. *American Economic Review*, 64, 291–303.

Krueger, A. (1985) The importance of general policies to promote economic growth. *World Economy*, 8, 93–108.

Krugman, P. (1984) Comments on Díaz-Alejandro. *Brookings Papers on Economic Activity*, 2, 390–393.

Kuczynski, P.-P. (1988) *Latin American Debt*. Baltimore: Johns Hopkins University Press.

Laird, S. and Nogués, J. (1989) Trade policies and highly indebted countries. *World Bank Economic Review*, 3, 241–61.

Lal, D. (1983a) *The Poverty of 'Development Economics'*. London: Institute of Economic Affairs.

Lal, D. (1983b) Time to put the Third World debt threat into perspective. *The Times*, 6 May.

Lehmann, D. (1990) *Democracy and Development in Latin America: Economics, Politics and Religion in the Postwar Period*. Cambridge: Polity.

Lenin, V. (1970) *Imperialism: the Highest Stage of Capitalism*. Peking: Foreign Languages Press.

Lernoux, P. (1986) *In Banks We Trust*. Harmondsworth: Penguin.

Lessard, D. and Williamson, J. (1987) *Capital Flight and Third World Debt*. Washington, DC: Institute for International Economics.

Lever, H. and Huhne, C. (1987) *Debt and Danger: the World Financial Crisis*. Harmondsworth: Pelican.

Lewis, M. (1989) *Liar's Poker*. New York: W. W. Norton.

Lewis, W. A. (1955) *The Theory of Economic Growth*. London: George Allen and Unwin.

Liebenstein, H. (1957) *Economic Backwardness and Economic Growth*. London: John Wiley.

Lindert, P. (1989) Response to debt crisis: what is different about the 1980s? In B. Eichengreen and P. Lindert (eds) *The International Debt Crisis in Historical Perspective*. Cambridge, MA: MIT Press, 227–75.

Lipietz, A. (1982) Towards Global Fordism? Marx or Rostow? *New Left Review*, 132, 33–58.

Lipietz, A. (1987) *Mirages and Miracles: the Crises of Global Fordism*. London: Verso.

Lipietz, A. (1989) The debt problem, European integration and the new phase of world crisis. *New Left Review*, 178, 37–56.

Lipsey, R. (1975) *An Introduction to Positive Economics*, 4th edn. London: Weidenfeld and Nicolson.

Little, I. M. (1982) *Economic Development*. New York: Basic Books.

Lustig, N. (1990) Economic crisis, adjustment and living standards in Mexico, 1982–1985. *World Development*, 18, 1325–42.

McCloskey, D. (1986) *The Rhetoric of Economics*. Madison: University of Wisconsin Press.

MacEwan, A. (1986) Latin America: why not default? *Monthly Review*, 38, 1–13.

Maddison, A. (1985) *Two Crises: Latin America and Asia, 1928–38 and 1973–83*. Paris: OECD.

Makin, J. (1984) *The Global Debt Crisis: America's Growing Involvement*. New York: Basic Books.

Mandel, E. (1975) *Late Capitalism*. London: Verso.

Marcel, M. and Palma, G. (1988) Third World debt and its effects on the British economy. *Cambridge Journal of Economics*, 12, 361–400.

Marglin, S. and Schor, J. (eds) (1990) *The Golden Age of Capitalism: Re-interpreting the Post-war Experience*. Oxford: Clarendon.

Marx, K. (1974) The critique of the Gotha Programme. In D. Fernbach (ed.), *Karl Marx: the First International and After*. Harmondsworth: Penguin, 339–59.

Marx, K. (1976) *Capital (Volume 1)*. Harmondsworth: Penguin.

Marx, K. and Engels, F. (1967) *The Communist Manifesto*. Harmondsworth: Penguin.

Milivojevic, M. (1985) *The Debt Rescheduling Process*. New York: St Martin's Press.

Miller, D. (1991) John Rawls. In D. Miller (ed.) *The Blackwell Encyclopaedia of Political Thought*. Oxford: Blackwell, 422–3.

Minogue, K. (1990) Equality: a response. In G. Hunt (ed.), *Philosophy and Politics*. Cambridge: Cambridge University Press, 99–108.

Morgan Guaranty Trust (1983) Global debt: assessment and prescriptions. *World Financial Markets*, February.

Morgan Guaranty Trust (1984) The LDC debt problem – at the midpoint? *World Financial Markets*, October–November.

Morgan Guaranty Trust (1986) The Baker initiative: the perspective of the banks. *World Financial Markets*, February.

Muntemba, D. (1989) The impact of the IMF–World Bank programmes on women and children in Zambia. In B. Onimode (ed.) *The IMF, the World Bank and the African Debt: Volume 2 – the Social and Political Impact*. London: Zed, 111–24.

Nelson, J. (ed.) (1990) *Economic Crisis and Policy Choice: the Politics of Adjustment in the Third World*. Princeton, NJ: Princeton University Press.

Nove, A. (1983) *The Economics of Feasible Socialism*. London: George Allen and Unwin.

Nozick, R. (1974) *Anarchy, State and Utopia*. Oxford: Blackwell.

Nunnenkamp, P. (1986) *The International Debt Crisis of the Third World*. Brighton: Wheatsheaf.

Nurske, R. (1953) *Problems of Capital Formation in Underdeveloped Countries*. Oxford: Blackwell.

O'Connell, A. (1985) External debt and the reform of the international monetary system. *Cepal Review*, 30, 51–66.

Ollman, B. (1976) *Alienation: Marx's Conception of Man in Capitalist Society*. Cambridge: Cambridge University Press.

Onimode, B. (ed.) (1989a) *The IMF, the World Bank and the African Debt: Volume 1 – The Economic Impact*. London: Zed.

Onimode, B. (ed.) (1989b) *The IMF, the World Bank and the African Debt: Volume 2 – The Social and Political Impact*. London: Zed.

Osiatyński, J. (ed.) (1990a) *Collected Works of Michal Kalecki: Volume 1 – Capitalism: Business Cycles and Full Employment*. Oxford: Oxford University Press.

Osiatyński, J. (ed.) (1990b) *Collected Works of Michal Kalecki: Volume 2 – Capitalism: Economic Dynamics*. Oxford: Oxford University Press.

Palma, G. (1978) Dependency: a formal theory of underdevelopment or a methodology for the analysis of concrete situations of underdevelopment? *World Development*, 6, 881–924.

Palma, G. (1989) Dependency. In J. Eatwell, M. Milgate and P. Newman (eds), *The New Palgrave: Economic Development*. London: Macmillan, 91–7.

Palmer, J. (1983) The debt-bomb threat. *Time Magazine*, 10 January.

Parboni, R. (1981) *The Dollar and Its Rivals: Recession, Inflation and International Finance*. London: Verso.

Parfitt, T. and Riley, S. (1988) African debt and the debate on stabilisation: the case of Nigeria. *Geoforum*, 19, 93–108.

Payer, C. (1974) *The Debt Trap: the IMF and the Third World*. London: Monthly Review Press.

Payer, C. (1991) *Lent and Lost: Foreign Credit and Third World Development*. London: Zed.

Peet, R. (1987) Industrial devolution, underconsumption and the Third World debt crisis. *World Development*, 15, 777–88.

Prebisch, R. (1951) The spread of technical progress and the terms of trade. *UN Economic Survey of Latin America 1949*, 46–61.

Prebisch, R. (1982) A historical turning point for the Latin American periphery. *Cepal Review*, 18, 7–24.

Raffer, K. (1989) International debts: a crisis for whom? In H. Singer and S. Sharma (eds), *Economic Development and World Debt*. London: Macmillan, 51–62.

Rawls, J. (1972) *A Theory of Justice*. Oxford: Oxford University Press.

Roddick, J. (1988) *The Dance of the Millions: Latin America and the Debt Crisis*. London: Latin America Bureau.

Rodrik, D. (1990) How should structural adjustment programs be designed? *World Development*, 18, 933–47.

Roett, R. (1989) How the 'haves' manage the 'have-nots': Latin America and the debt crisis. In B. Stallings and R. Kaufman (eds) *Debt and Democracy in Latin America*. Boulder, CO: Westview, 59–73.

Roemer, J. (1981) *Analytical Foundations of Marxian Economic Theory*. Cambridge: Cambridge University Press.

Roemer, J. (1988) *Free to Lose.* Cambridge, MA: Harvard University Press.

Rohatyn, F. (1983) A plan for stretching out global debt. *Business Week,* 28 February.

Rosenstein-Radan, P. N. (1943) Problems of industrialization of Eastern and South-Eastern Europe. *Economic Journal,* 53, 202–11.

Rostow, W. W. (1960) *The Stages of Economic Growth: a Non-communist Manifesto.* London: Cambridge University Press.

Rowthorn, R. (1980) *Capitalism, Conflict and Inflation.* London: Lawrence and Wishart.

Roxborough, I. (1989) Organized labor: a major victim of the debt crisis. In B. Stallings and R. Kaufman (eds), *Debt and Democracy in Latin America.* Boulder, CO: Westview Press, 91–108.

Sachs, J. (1984) Comments on Díaz-Alejandro. *Brookings Papers on Economic Activity,* 2, 393–401.

Sachs, J. (1985) External debt and macroeconomic performance in Latin America and East Asia. *Brookings Papers on Economic Activity,* 2, 523–74.

Sachs, J. (1986) The Bolivian hyperinflation and stabilization. NBER Working Paper 2073. Cambridge, MA: National Bureau of Economic Research.

Sachs, J. (1987) Trade and exchange rate policies in growth-oriented adjustment programs. In V. Corbo, M. Goldstein and M. Khan (eds), *Growth-oriented Adjustment Programs.* Washington, DC: IMF and the World Bank, 291–325.

Sachs, J. (1988) The debt crisis at a turning point. *Challenge,* May–June.

Sachs, J. (ed.) (1989a) *Developing Country Debt and the World Economy.* Chicago: University of Chicago Press/NBER.

Sachs, J. (1989b) Introduction. In J. Sachs (ed.), *Developing Country Debt and the World Economy.* Chicago: University of Chicago Press/NBER, 1–33.

Sachs, J. and Huizinga, H. (1987) The US commercial banks and the developing-country debt crisis. *Brookings Papers on Economic Activity,* 2, 555–601.

Sayer, D. (1991) *Capitalism and Modernity: an Excursus on Marx and Weber.* London: Routledge.

Schultz, T. (1964) *Transforming Traditional Agriculture.* New Haven, CT: Yale University Press.

Schultz, T. (1987) Tensions between economics and politics in dealing with agriculture. In G. Meier (ed.), *Pioneers in Development (Second Series).* Oxford: Oxford University Press/World Bank, 17–38.

Selowsky, M. and van der Tak, H. (1986) The debt problem and growth. *World Development*, 14, 1107–24.

Seth, R. and McCauley, R. (1987) Financial consequences of new Asian surpluses. *Federal Reserve Bank of New York Quarterly Review*, 12, 32–44.

Simmel, G. (1990) *The Philosophy of Money*. London: Routledge.

Singh, A. (1988) Employment and output in a semi-industrial country: modelling alternative policy options in Mexico. In M. Hopkins (ed.) *Employment Forecasting*. London: Pinter, 184–209.

Sjaastad, L. (1983) International debt quagmire – to whom do we owe it? *The World Economy*, 6, 305–24.

Smith, D. (1986) Marxian economics. In R. Johnston (ed.), *Dictionary of Human Geography*. Oxford: Blackwell, 282–7.

Smith, N. (1984) *Uneven Development: Nature, Capital and the Production of Space*. Oxford: Blackwell.

Stallings, B. (1987) *Banker to the Third World: US Portfolio Investment in Latin America, 1900–1986*. Berkeley: University of California Press.

Stewart, F. (1984) The fragile foundations of the neo-classical approach to development. *Journal of Development Studies*, 21, 282–92.

Stewart, F. (1985) The international debt situation and North–South relations. *World Development*, 13, 191–204.

Strange, S. (1986) *Casino Capitalism*. Oxford: Blackwell.

Streeten. P. (1987) Structural adjustment: a survey of the issues and options. *World Development*, 15, 1469–82.

Tavares, M. (1985) The revival of American hegemony. *Cepal Review*, 23, 139–46.

Taylor, L. (1987) *Varieties of Stabilization Experience: Towards Sensible Macroeconomics in the Third World*. New York: Oxford University Press.

Thrift, N. and Leyshon, A. (1988) 'The gambling propensity': banks, developing country debt exposures and the new international financial system. *Geoforum*, 19, 55–69.

Tokman, V. (1985) Global monetarism and destruction of industry. *Cepal Review*, 23, 107–21.

Tomassini, L. (1984) The international scene and Latin America's external debt. *Cepal Review*, 24, 135–46.

Toye, J. (1983) The disparaging of development economics. *Journal of Development Studies*, 20, 87–107.

Toye, J. (1987) *Dilemmas of Development: Reflections on the Counter-revolution in Development Theory and Policy*. Oxford: Blackwell.

Triffin, R. (1960) *Gold and the Dollar Crisis*. New Haven, CT: Yale University Press.

UNCTAD (1986) *Trade and Development Report, 1985.* Geneva: United Nations.

Vaubel, R. (1983) The moral hazard of IMF lending. *The World Economy*, 6, 291–303.

Vaughn, K. (1989) The invisible hand. In J. Eatwell, M. Milgate and P. Newman (eds), *The New Palgrave: The Invisible Hand.* London: Macmillan, 168–72.

Volcker, P. (1983) Remarks before the Subcommittee on International Finance and Monetary Policy of the Committee on Banking, Housing, and Urban Affairs, US Senate, February 17th. *Federal Reserve Bulletin*, 69, 175–7.

Wachtel, H. (1986) *The Money Mandarins: the Making of a Supranational Economic Order.* New York: Pantheon Books.

Wade, R. (1984) Dirigisme Taiwan-style. *IDS Bulletin*, 15, 65–70.

Wallich, H. (1984) Insurance of bank lending to developing countries. Group of Thirty Occasional Paper, 15.

Walton, J. (1989) Debt, protest and the state in Latin America. In S.´ Eckstein (ed.), *Power and Popular Protest: Latin American Social Movements.* Berkeley: University of California Press, 299–328.

Warren, B. (1980) *Imperialism: Pioneer of Capitalism.* London: Verso.

Watts, M. (1989) The agrarian question in Africa: debating the crisis. *Progress in Human Geography*, 13, 1–41.

Weber, M. (1978) *Economy and Society.* Berkeley: University of California Press.

Wolfe, T. (1987) *The Bonfire of the Vanities.* London: Picador.

Wolff, R. and Resnick, S. (1987) *Economics: Marxian versus Neo-classical.* Baltimore: Johns Hopkins University Press.

Woo, W. T. and Nasution, A. (1989) The conduct of economic policies in Indonesia and its impact on external debt. In J. Sachs (ed.), *Developing Country Debt and the World Economy.* Chicago: University of Chicago Press/NBER, 101–20.

Wood, B. M. (1989) Rational choice Marxism: is the game worth the candle? *New Left Review*, 177, 41–88.

Wood, R. (1986) *From Marshall Plan to Debt Crisis.* Berkeley: University of California Press.

World Bank (1978) *World Development Report, 1978.* Oxford: Oxford University Press/World Bank.

World Bank (1981) *Accelerated Development in Sub-Saharan Africa.* Washington, DC: The World Bank.

World Bank (1982a)–(1991a) *World Debt Tables (Annual).* Washington, DC: The World Bank.

World Bank (1982b)–(1991b) *World Development Reports (Annual).*

Washington, DC: The World Bank.

Wriston, W. (1986) *Risk and Other Four Letter Words*. New York: Harper and Row.

Yeats, A. (1990) On the accuracy of economic observations: do sub-Saharan trade statistics mean anything? *World Bank Economic Review*, 4, 135–56.

Index